COMPUTER-AIDED DESIGN AND VLSI DEVICE DEVELOPMENT

**THE KLUWER INTERNATIONAL SERIES
IN ENGINEERING AND COMPUTER SCIENCE**

VLSI, COMPUTER ARCHITECTURE AND
DIGITAL SIGNAL PROCESSING

Consulting Editor

Jonathan Allen

COMPUTER-AIDED DESIGN AND VLSI DEVICE DEVELOPMENT

KIT MAN CHAM
Hewlett-Packard Laboratories

SOO-YOUNG OH
Hewlett-Packard Laboratories

DAEJE CHIN
IBM—Thomas J. Watson Research Center

JOHN L. MOLL
Hewlett-Packard Laboratories

KLUWER ACADEMIC PUBLISHERS
Boston/Dordrecht/Lancaster

Distributors for North America:
Kluwer Academic Publishers
190 Old Derby Street
Hingham, Massachusetts 02043, USA

Distributors for the UK and Ireland:
Kluwer Academic Publishers
MTP Press Limited
Falcon House, Queen Square
Lancaster LA1 1RN, UNITED KINGDOM

Distributors for all other countries:
Kluwer Academic Publishers Group
Distribution Centre
Post Office Box 322
3300 AH Dordrecht, THE NETHERLANDS

Library of Congress Cataloging-in-Publication Data
Main entry under title:

Computer-aided design and VLSI device development.

   (The Kluwer international series in engineering
and computer science ; SECS   . VLSI, computer
architecture, and digital signal processing)
   Bibliography: p.
   Includes index.
   1. Integrated circuits—Very large scale integration—
Design and construction—Data processing. 2. Computer-
aided design. I. Cham, Kit Man. II. Series: Kluwer
international series in engineering and computer
science ; SECS   . III. Series: Kluwer international
series in engineering and computer science. VLSI,
computer architecture, and digital signal processing.
TX7874.C6474 1986   621.395   85–19916

ISBN-13: 978-1-4612-9605-8   e-ISBN-13: 978-1-4613-2553-6
DOI:10.1007/ 978-1-4613-2553-6

CONTENTS

PREFACE   ix

OVERVIEW   1

PART A: SIMULATION SYSTEMS   11

CHAPTER 1. INTRODUCTION TO NUMERICAL SIMULATION
          SYSTEM   13

  1.1 HISTORY OF NUMERICAL PROCESS AND DEVICE
     SIMULATIONS   13
  1.2 IMPLEMENTATION OF NUMERICAL SIMULATION
     SYSTEM   15

CHAPTER 2. PROCESS SIMULATION   23

  2.1 INTRODUCTION   23
  2.2 SUPREM : 1-D PROCESS SIMULATOR   25
  2.3 SUPRA : 2-D PROCESS SIMULATOR   40
  2.4 SOAP : 2-D OXIDATION SIMULATOR   53

CHAPTER 3. DEVICE SIMULATION   65

  3.1 GEMINI : 2-D POISSON SOLVER   65
  3.2 CADDET : 2-D 1-CARRIER DEVICE SIMULATOR   82
  3.3 SIFCOD : GENERAL-SHAPE 2-D 2-CARRIER
     DEVICE SIMULATOR   96

CHAPTER 4. FCAP2 : PARASITIC CAPACITANCE/RESISTANCE
          SIMULATOR   113

  4.1 INTRODUCTION   113
  4.2 INPUT AND FLOWCHART OF FCAP2   115
  4.3 EXAMPLE   117

PART B: APPLICATIONS AND CASE STUDIES   121

CHAPTER 5. METHODOLOGY IN COMPUTER-AIDED DESIGN FOR
          PROCESS AND DEVICE DEVELOPMENT   123

  5.1 METHODOLOGIES IN DEVICE SIMULATIONS   123
  5.2 OUTLINE OF THE CASE STUDIES   129

CHAPTER 6. BASIC TECHNIQUES IN SIMULATIONS FOR
          ADVANCED PROCESS DEVELOPMENT   131

  6.1 BASIC DEVICE PHYSICS FOR PROCESS
     DEVELOPMENT   131
  6.2 CAD TOOLS FOR SIMULATION OF DEVICE
     PARAMETERS   139

6.3 METHODS OF GENERATING BASIC DEVICE
    PARAMETERS   144
6.4 RELATIONSHIP BETWEEN DEVICE CHARACTERISTICS
    AND PROCESS PARAMETERS   150

CHAPTER 7. DRAIN-INDUCED BARRIER LOWERING IN SHORT
           CHANNEL TRANSISTORS   159

CHAPTER 8. TRANSISTOR DESIGN FOR SUBMICRON CMOS
           TECHNOLOGY   171

8.1 INTRODUCTION TO SUBMICRON CMOS TECHNOLOGY   171
8.2 DEVELOPMENT OF THE SUBMICRON P-CHANNEL
    MOSFET USING SIMULATIONS   176
8.3 N-CHANNEL TRANSISTOR SIMULATIONS   191
8.4 SUMMARY   196

CHAPTER 9. THE SURFACE INVERSION PROBLEM IN TRENCH
           ISOLATED CMOS   199

9.1 INTRODUCTION TO TRENCH ISOLATION IN
    CMOS   199
9.2 SIMULATION TECHNIQUES   201
9.3 ANALYSIS OF THE INVERSION PROBLEM   205
9.4 SUMMARY OF SIMULATION RESULTS   211
9.5 EXPERIMENTAL RESULTS   212
9.6 SUMMARY   214

CHAPTER 10. DEVELOPMENT OF ISOLATION STRUCTURES FOR
            APPLICATIONS IN VLSI   217

10.1 INTRODUCTION TO ISOLATION STRUCTURES   217
10.2 LOCAL OXIDATION OF SILICON (LOCOS)   219
10.3 MODIFIED LOCOS   223
10.4 SIDE WALL MASKED ISOLATION (SWAMI)   228
10.5 SUMMARY   234

CHAPTER 11. A STUDY OF LDD DEVICE STRUCTURE USING 2-D
            SIMULATIONS   239

11.1 HIGH ELECTRIC FIELD PROBLEM IN
     SUBMICRON MOS DEVICES   239
11.2 LDD DEVICE STUDY   241
11.3 SUMMARY   258

CHAPTER 12. MOSFET SCALING BY CADDET   261

12.1 INTRODUCTION   261
12.2 SCALING OF AN ENHANCEMENT MODE
     MOSFET   262
12.3 SCALING OF A DEPLETION MODE MOSFET   271
12.4 CONCLUSIONS   281

CHAPTER 13. PARASITICS EXTRACTION FOR VLSI PROCESS
               DEVELOPMENT   283

   13.1 INTRODUCTION   283
   13.2 SIMULATION TECHNIQUES   285
   13.3 DIFFUSION CAPACITANCE BY SUPRA
         SIMULATIONS   297
   13.4 PARASITIC RESISTANCE BY FCAP2 SIMULA-
         TIONS   298
   13.5 EXPERIMENT AND SIMULATION COM-
         PARISONS   298
   13.6 SUMMARY   299

APPENDIX   301

TABLE OF SYMBOLS   303

SUBJECT INDEX   309

ABOUT THE AUTHORS   315

PREFACE

This book is concerned with the use of Computer-Aided Design (CAD) in the device and process development of Very-Large-Scale-Integrated Circuits (VLSI). The emphasis is in Metal-Oxide-Semiconductor (MOS) technology. State-of-the-art device and process development are presented.

This book is intended as a reference for engineers involved in VLSI development who have to solve many device and process problems. CAD specialists will also find this book useful since it discusses the organization of the simulation system, and also presents many case studies where the user applies the CAD tools in different situations. This book is also intended as a text or reference for graduate students in the field of integrated circuit fabrication. Major areas of device physics and processing are described and illustrated with simulations.

The material in this book is a result of several years of work on the implementation of the simulation system, the refinement of physical models in the simulation programs, and the application of the programs to many cases of device developments. The text began as publications in journals and conference proceedings, as well as lecture notes for a Hewlett-Packard internal CAD course.

This book consists of two parts. It begins with an overview of the status of CAD in VLSI, which points out why CAD is essential in VLSI development. Part A presents the organization of the two-dimensional simulation system. The process, device and parasitics simulation programs are described in some detail. The basic principles, input file format and application examples are presented. These chapters are intended to introduce the reader to the programs. Since these programs are in the public domain, the reader is referred to the manuals for more details. Part B of the book presents case studies, where the application of simulation tools to solve VLSI device design problems is described in detail. The physics of the problems are illustrated with the aid of numerical simulations. Solutions to these problems are presented. Issues in state-of-the-art device development such as drain-induced barrier lowering, trench isolation, hot electron effects, device scaling and interconnect parasitics are discussed.

For the book to be used as a textbook, we recommend that it be used for a semester course. If the course deals with device modeling and computer-aided design tools, then Part A of the book should be emphasized. The student will learn about the fundamentals of process and device simulation programs and simulation system organization. If the course deals with device physics and process development, then Part B of the book should be emphasized. The student will learn about current issues in VLSI device development. In either case, Part A and Part B of the book will complement each other.

We are grateful that Dr. John Chi-Hung Hui has contributed Chapter 11 on the issue of hot electron degradation effects in submicron n-channel MOSFETs. The optimization of the LDD structure for reducing the hot electron degradation is described in detail.

We are also grateful that Dr. Sukgi Choi has contributed Chapter 12 on the issue of device scaling. The scaling of n-channel enhancement and depletion mode devices are presented. Factors causing the device characteristics to deviate from classical scaling rules, as well as complications involved in short channel device scaling such as punchthrough are discussed.

We are indebted to Dr. P. Vande Voorde, Dr. D. Wenocur and Mr. M. Varon for proofreading the manuscripts. Mr. K. Ogasaki has been assisting us with the computer graphics which produced many of the figures in this book. Thanks are to Mr. T. Ekstedt who has kindly assisted the formatting of the text, and Dr. S.-L. Ng who has assisted in preparing many of the figures.

The manuscript was prepared by the TDP software on the HP3000 computer. Most of the simulations were performed on the HP1000 computer. We are grateful to the system managers for their support and assistance.

We are indebted to many of our colleagues at Hewlett-Packard Laboratories for providing us with many ideas and suggestions, and to our management for providing an opportunity for us to complete this task.

Finally, we would like to thank our families for their spiritual support, patience and understanding during preparation of the manuscript.

COMPUTER-AIDED DESIGN AND VLSI DEVICE DEVELOPMENT

Overview

In order to bring out the importance of Computer-Aided Design(CAD) in VLSI(Very-Large-Scale Integration) device design, it is necessary to discuss the evolution of the Metal-Oxide-Semiconductor Field-Effect Transistors (MOSFETs) and the issues involved in its scaling. MOSFETs, first proposed 50 years ago, are based on the principle of modulating longitudinal electrical conductance by varying a transverse electrical field. Since its conception, MOSFET technology has improved steadily and has become the primary technology for large-scale circuit integration on a monolithic chip, primarily because of the simple device structure. VLSI development for greater functional complexity and circuit performance on a single chip is strongly motivated by the reduced cost per device and has been achieved in part by larger chip areas, but predominantly by smaller device dimensions and the clever design of devices and circuits.

A general guide to the smaller devices in MOSFETs and associated benefits, has been proposed by Dennard et al [1] (MOSFET scaling). This proposed scheme assumes that the x and y dimensions (in the circuit plane) are large compared to the z-dimension for the active device. The scaling method is also restricted to MOS devices and circuits. The active portion of MOS devices is typically restricted to within one or two microns of the crystal surface and interconnection dielectrics and metals are less than one micron in thickness. Thus we should expect that the guidelines as proposed by Dennard should be reasonably valid for minimum circuit dimensions of two microns or greater. As the dimensions decrease below two microns, problems are introduced in both

1

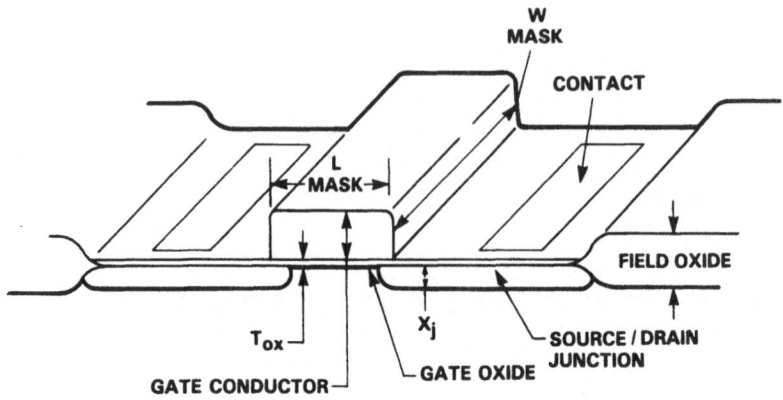

Fig. 1 Generic MOS Transistor.

fabrications and device operations that are not significant in larger long-channel devices. The 2-D aspects of the impurity profiles and oxidation process become important in determining the effective channel length and width. More processing steps are required, such as channel implantation and local oxidation, which make more stringent control of the process necessary. Secondary effects, such as oxidation-enhanced diffusion significantly affect the impurity profile. As a result, better understanding and accurate control of these phenomena are crucial to achieving the desired performance from the scaled devices.

For the device operation, we will examine many of the scaling assumptions as applied to the long channel, wide conductor circuits. Also, in each case, there have been practical departures from the scaling assumptions. The original rules proposed that physical dimensions were scaled so that all electric field patterns were kept constant. Fig. 1 shows a generic MOS transistor and the various dimensions. Table 1 gives Dennard's constant field scaling rule, even though these have not been the general practice. When a process is scaled to smaller dimensions, the x,y and z dimensions are all scaled by the same amount. In addition, the applied voltages are scaled in constant field scaling to maintain constant field pattern.

$$W, L, T_{ox}, V, N_b \qquad \propto K^{-1}$$

$$I_{DS} \quad \propto (W/L)(V^2/T_{ox}) \propto K^{-1}$$

$$C_g \quad \propto W L C_{ox} \qquad \propto K^{-1}$$

$$t_d \quad \propto C_g V / I \qquad \propto K^{-1}$$

$$P \quad \propto V I \qquad \propto K^{-2}$$

$$P/A \quad \propto V I / W L \qquad \propto 1$$

$$P t_d \qquad \propto K^{-3}$$

Table 1 Dennard's constant field scaling.

Many scaling schemes have been proposed since the "constant field" proposal. It is useful to consider the actual scaling methods that have been followed. The desirability of electrical compatibility with bipolar TTL circuits and the five volt power supply standard has resulted in "constant voltage scaling" as far as circuit power supply is concerned. Sometimes the internal node voltages are changed as a result of scaling and re-design. If the source/drain junction is less than one micron and V_{DD} is retained at five volts, the electric field stress in the channel is too great and the MOS transistor characteristics drift with time because of the hot carrier charge trapping. It is possible, but not evident at this writing, that some modified structure such as a graded junction or a "low doped drain" (LDD) can allow five volt operation for sub-micron devices. This constant voltage scaling also makes the 2-D field coupling significant, which is negligible in the long-channel device. It is the major cause of all the short-channel effects. To model these short-channel effects, 2-D numerical simulations become necessary because 1-D analytical models are not adequate. We must, in any event, consider lower system voltage at some future time. Following examples illustrate several features of the scaling methods. The most evident is that whereas most features of Dennard's constant field scaling are approximately retained from one generation of technology to the next, practical considerations have resulted in significant departures. Constant voltage scaling has in fact been the primary mode of scaling for most merchant suppliers. We can expect that a new power supply standard will be adopted and used until the dimensions are once more so small that device instability re-appears. Another departure from strict geometrical scaling has been in the vertical thickness of films. Conductor thickness has scaled very slowly in order to avoid electromigration effects in aluminum, or signal delay effects in polysilicon conductors. Table 2 shows the actual scaling done by most industrial suppliers. Either constant voltage or constant field scaling has resulted in improved circuits as measured by speed, and chip size and power for a given electronic function. As was mentioned earlier, circuit voltage, V_{DD}, will almost certainly be reduced for sub-micron devices. The tendency not to scale the interconnect or dielectric thickness (except gate oxide) will continue.

The future reduction of minimum features to less than one micron will undoubtedly bring further changes in actual scaling effects. The minimum practical conduction threshold for switches to turn off in dynamic circuits is approximately 0.6 volts. The desirability of dynamic operation in many electrical functions will keep CMOS circuit operation at about two volts. There will be exceptions such as the watch circuit that operates from a single battery cell. The peak current ($V_{GS} = V_{DS}/2$) per unit width scales as K^{-2} for long channel devices and constant voltage scaling. The effect of velocity saturation is to reduce this scaling factor to approximately K^{-1}. If width is also reduced by the same scaling factor of K, then peak current per unit width scales as K^{-1} for long channel and is almost constant for short channel. The transconductance follows the same behavior as current. Some switches such as the transistors in a static RAM cell are not required to switch particularly fast and have a small capacitive load. Hence the relative lack of increased performance from scaling, the smallest geometry device does not result in a performance problem inside the static RAM cell. On

PROCESS (COMPANY)	Leff (um)	Vt (volt)	VDD (volt)	Tox (nm)	Xj (um)
NMOS(Intel)	4.6		5.0	120	2.00
NMOS(HP)	3.0	0.8	5.0	100	
HMOSI(Intel)	2.9	0.7	5.0	70	0.80
NMOS(Xerox)	2.5		5.0	70	0.46
NMOS(HP)	2.0	0.8	5.0	50	0.20
HMOSII(Intel)	1.6	0.7	5.0	40	0.80
NMOS(HP)	1.4	0.6	3.0	40	0.30
HMOSIII(Intel)	1.1	0.7	5.0	25	0.30
NMOS(NTT)	1.0	0.5	5.0	30	0.25
NMOS(IBM)	0.8	0.6	2.5	25	0.35
NMOS(Toshiba)	0.5	0.5	3.0	15	0.23
NMOS(AT&T)	0.3		1.5	25	0.26

Table 2 Actual scaling done by the most industrial suppliers.

the other hand, devices that must drive long signal lines, clock lines, or word lines may require width-scaling that is not at all the same as length-scaling.

The scaling for commercial devices then will be to a new voltage standard of less than five volts as dimensions become sub-micron. The tendency in node capacitance is such that average wire length is a fraction of chip size. As more electronic functions are included on a chip, the chip size will continue to increase, and wire length also increases. In future scaling, the capacitance per unit length of wire for minimum pitch will stay almost constant since a large part of capacitance is fringing field or else inter-line. A new circuit design problem will be wire placement to minimize capacitance effects. To evaluate the actual capacitance values, the circuit extraction program coupled with the 2-D or 3-D parasitics simulator is indispensable.

An additional feature of scaling will be that the logic device width will be reduced by a smaller factor than the scaling factor. This feature is a result of the fact that the driving current per unit width only scales at best as $C_0 \sim K^{-1}$. If the width is scaled by a factor of K, then the driving current per device at constant voltage will decrease somewhat as a result of various parasitic effects. The examples of minimum operating voltage and deleterious effects on performance are given to help illustrate the practical approach to sub-micron scaling.

The ability to calculate the effect of a process change on circuit or device electrical parameters has been an indispensable part of the rapid advances that have been made in semiconductor circuits ever since the beginning of the solid state micro-electronic industry. Experiments tend to establish the validity of theoretical concepts, establish empirical laws, as well as help discover new physical effects. A symbiontic relation has developed between experiment and associated theory (modeling) in which the modeling helps to guide the direction of experiments, and experiments establish the validity of models as well as produce devices and circuits with optimized performance.

The conventional process and device designs for integrated-circuit technologies have been based on a trial-and-error approach using

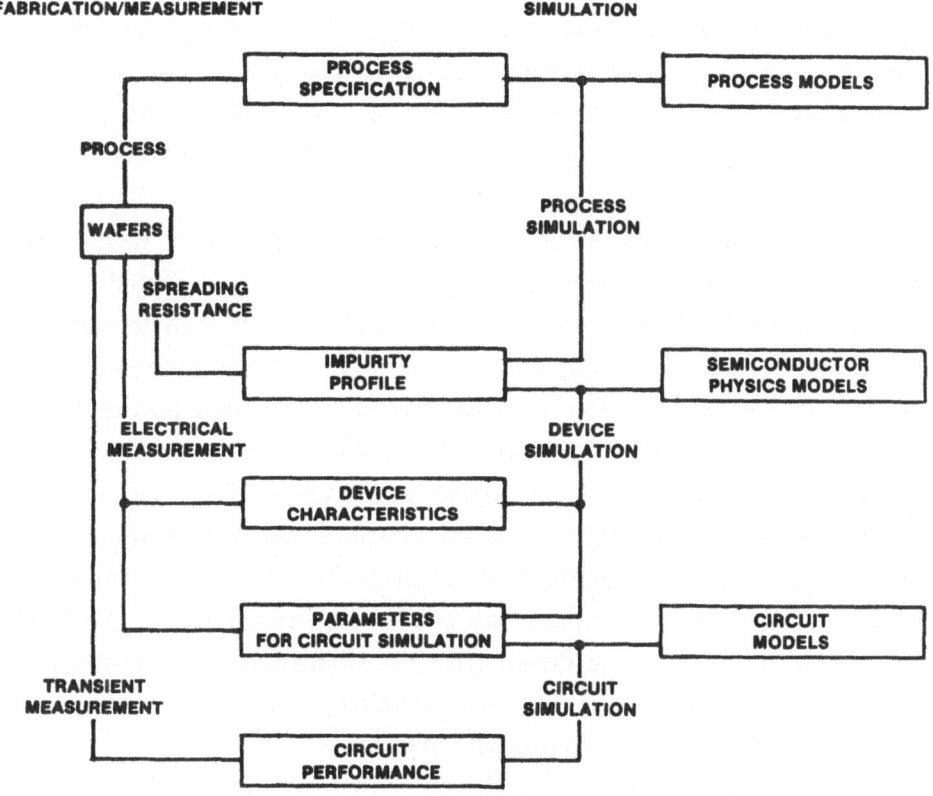

Fig. 2 Block diagram of the process development

fabrications and measurements plus simple 1-D analytical modeling to achieve the desired terminal electrical characteristics and circuit performances. The left half of Fig. 2 outlines a process, device, and circuit design using the fabrications and simple models. This approach is not, however, adequate for the small geometry devices. As mentioned in the review of scaling, the constant voltage scaling and the lack of vertical scaling make the 2-D field coupling more dominant in the device performance. Especially, the threshold voltage becomes a sensitive function of the channel length and the drain bias. The fringing and inter-line capacitances become significant in the wiring capacitance. The velocity saturation also prohibits the simple 1-D model from accurately predicting the saturation currents. These factors force the engineers to resort more to the experiments. Thus, it drastically increases the cost and time to develop a scaled geometry process. Even with the experiments, complicated processes and structures make it difficult to get physical insight and quantitative analysis of the factors governing device operation.

A complementary analysis and design path through process, device, and circuit simulations has been proposed and is now widely accepted. In the process simulation, process-specification information is used to simulate the device structure and impurity distribution using the process models. Device simulation yields the terminal characteristics based on the device structure and the impurity profile from the process simulation and physical models. The SPICE parameters are extracted from the terminal characteristics and the layouts. Based on these parameters and circuit connectivity, circuit simulations yield the switching characteristics and provides the means to evaluate the circuit performances. Compared to laboratory experiment, the design path via simulation is less costly and faster; more important, it produces detailed information concerning device operation in a well-controlled environment.

A complete 2-D numerical simulation system has been implemented in Hewlett-Packard Laboratories since 1982. In part A, this numerical simulation system and its individual tools will be explained in detail so that the reader can get acquainted with these tools and learn how to use them. Most of these tools are in the public domain. Thus we also give the information of these programs in the appendix so that it will help the reader implement these tools. In part B, the applications of the system in modeling small geometry processes will be presented. These simulation tools are different with the simple analytical models. First, we try to develop a methodology to use these tools effectively in process development. Next, real examples which are typical in important topics of scaled process development will be given in detail to help the reader attack their real problems.

As can be seen through the examples, Computer-Aided-Design has been found to be absolutely essential in VLSI process and device development. As the structures on the integrated circuit, such as transistors, field isolations, and interconnects are scaled down in size, many issues which have not been important have become significant. In the case of the transistor, examples of these issues are the drain-induced barrier lowering, and hot electron effects. For the interconnects, the issues are fringing-field capacitance and interline coupling. The physics of these issues cannot be understood in simple terms. Numerical simulations are the best tool to solve these problems. Simulations have provided guidance in the device design, with little investment in time and cost.

Simulations have provided accurate values of some device parameters such as threshold voltage, and parasitic fringing-field capacitance. For some parameters, such as punchthrough voltage and subthreshold leakage, simulation is less accurate because it is dependent on many factors. The channel profile, all the way from the surface to the bulk, has to be accurately known. The source/drain profile will also affect the punchthrough calculations. Although even in these cases, where simulation is less accurate, they do provide good estimates on the trend. Also they provide better understanding of the physics of the problem. Only through simulations can one observe the internal potential of a device, or the electric field at the drain in the case of hot electron effects.

Reference

[1] R. H. Dennard, F. H. Gaensslen, H. N. Yu, V. L. Rideout, E. Bassous, and A. Le Blanc, "Design of Ion-Implanted MOSFET's with very Small Physical Dimensions," *IEEE J. Solid-State Circuits*, SC-9, Oct 1974, pp. 256-268.

Part A
Simulation Systems

Chapter 1

Introduction to Numerical Simulation System

1.1 History of Numerical Process and Device Simulations

Historically, numerical simulations of the MOS device have been tried first to understand the device operation in the subthreshold and saturation regions. In 1969, Barron [1.1] from Stanford University simulated a MOSFET transistor using a finite-difference method to study the subthreshold conduction and saturation mechanism. Vandorpe [1.2] also simulated and modeled the saturation region with the finite-difference program in 1972. After the self-aligned silicon gate technology was invented, the MOSFET device dimensions were reduced. This reduction prompted more numerical simulations to study the short-channel and narrow width effects. Mock and Kennedy [1.3] from IBM developed a finite-difference program. Hachtel [1.4] also from IBM developed the first finite-element device simulation program. Barnes [1.5] from University of Michigan also developed a finite-element device simulation program for GaAs MESFETs. Most of the programs mentioned above were developed as research tools rather than for the general user(design tools). More stress had been put on the development of a stable and fast algorithm and the implementations of the physical mechanisms rather than on the user interface.

As MOS devices have been shrunk further, several groups started to develop the 2-D device simulator as a general design tool for the small geometry devices. One group was Cottrell/Buturla [1.6] from IBM who developed the finite-element program FIELDAID. Mock [1.7] also

13

developed the finite-difference program CADDET using the stream function with Toyabe from Hitachi. Greenfield [1.8] and Dutton from Stanford University developed the finite-difference 2-D Poisson solver, GEMINI, which is only valid for the subthreshold and linear regions. Selberherr and Potzel [1.9] from University of Vienna developed 2-D 1-carrier program, MINIMOS using the finite-differnce method. With the exception of FIELDAID, these programs are limited in input geometries and simulate the steady-state case. They are sufficient for the conventional structure MOS devices. However, as the device structures become more complicated and the speed improves, it becomes necessary to have the capabilities to simulate the arbitrary-shape geometry and the transient phenomena. Recently, Mock [1.10] has developed the SIFCOD program to satisfy these needs. SIFCOD is a   2-D 2-carrier device simulator with the arbitrary-shape geometry and transient device simulations including the small number of lumped circuit elements. In 1984, Pinto [1.11] from Stanford University has also developed the PISCES II program, which is a 2-D 2-carrier device simulator with arbitrary-shape input.

Process simulation started much later compared with device simulation. In the 60's and 70's, most of process modelings had been done by the analytical equations such as the diffusion equation and Deal-Grove oxidation equation[1.12]. As the process became sophisticated and the device was scaled down, the secondary effects such as the doping dependence of diffusivity, oxidation enhanced diffusion, severely affected the accuracy of the simple analytical model. In 1977, Antoniadis and Dutton [1.13] from Stanford introduced the SUPREM program, which is a 1-D numerical process simulator. It became very popular and an indispensable tool in process development.

As the device shrinks further, the 2-D effects are increasingly important in the process modeling as in the device modeling. Lee [1.14] from Stanford University developed 2-D process simulator, BIRD, which employed Green's function to solve the diffusion equation. The Green's function approach requires constant diffusivity and is only valid in the low concentration case. To overcome this limitation, Chin and

Kump [1.15] also from Stanford University developed a numerical 2-D process simulator, SUPRA using the Green's function method for low impurity concentration as in BIRD and finite-difference method for high impurity concentration.

1.2 Implementation of Numerical Simulation System

As mentioned in the previous chapter, many 2-D simulation programs have been developed and several have been reported recently. Most of these programs are more research tools rather than design tools. More stress has been put on the development of fast algorithms and the implementation of the physical mechanism than on the user interface. Futhermore, each program was developed independently and without an interface to other programs. Thus, transferring massive amounts of 2-D numerical data from one program to another is very difficult. Analyzing and interpreting the data is also difficult. To overcome these problems and provide a convenient design path using simulation, a complete 2-D simulation system has been developed at Hewlett Packard with an emphasis on the user interface. The following schemes have been adopted to make this system a more practical, user-oriented and well supported design tool.

1) *Mode of program execution* : To reduce the engineering time, users want to control the input preparation, program execution and data file handling. Ideally, users can run the programs in interactive mode for the quick check of the input and the graphical post processing and run in batch mode for the time consuming simulation job. Thus, all the programs have been implemented on the mini-computer in our laboratory with an emphasis on the user control and fast turn-around.

2) *Hierarchical simulation* : Full approach for the 2-D simulations is time consuming. This is especially true in process and device simulations. Thus, simplified programs are implemented together with the full approach program so that users can select according to their application to save the simulation time.

3) *Interface between programs* : These simulation programs generate a large amount of 2-D data. A clean and user-transparent procedure is important to transfer data between the programs. Standard format of 2-D data file has been set up and utilized.

4) *Graphical post processing* : To manipulate and analyze the massive 2-D data graphical post processing is absolutely essential. A plot package has been implemented which can do the 3-D bird's-eye-view plot, 2-D contour plot, and 1-D cross-section plot all together.

5) *Complete system support* : For the better user acceptance, system support in the form of the training and manual is a critical issue. Users are usually afraid to read the manual and use programs without training because they may spend too much time without any results. Ideally, training can be a means for users to become acquainted with and begin to use a program. Thereafter, the manual can be used for reference. The manual should be easy to understand, complete and up-to-date. An on-line manual is very helpful.

6) *Easy to enhance* : The physical model in the process and device simulation is still evolving. Sometimes, the simulation programs are used as development vehicles for new models. Thus, it is necessary to make the programs modular so that they are easy to modify and repair.

7) *Bench-mark of the simulation tools* : Without bench-marking, the tools are useless. It is necessary either to bench-mark the tools in a wide range of the practical interest or specify the range where the tool is valid.

A block diagram of the system is shown in Fig. 1.1. For process simulation, there are the SUPREM, SUPRA and SOAP programs. SUPREM is a 1-D process program capable of simulating most typical IC fabrication steps such as deposition, etch, ion implantation, diffusion and oxidation. It generates the 1-D profile of all the dopants present in the silicon and silicon dioxide. SUPREM is ideal for simulation of the doping profile of the bipolar or MOS capacitor where the 1-D structure is dominant. The doping distribution of a MOS transistor is basically two

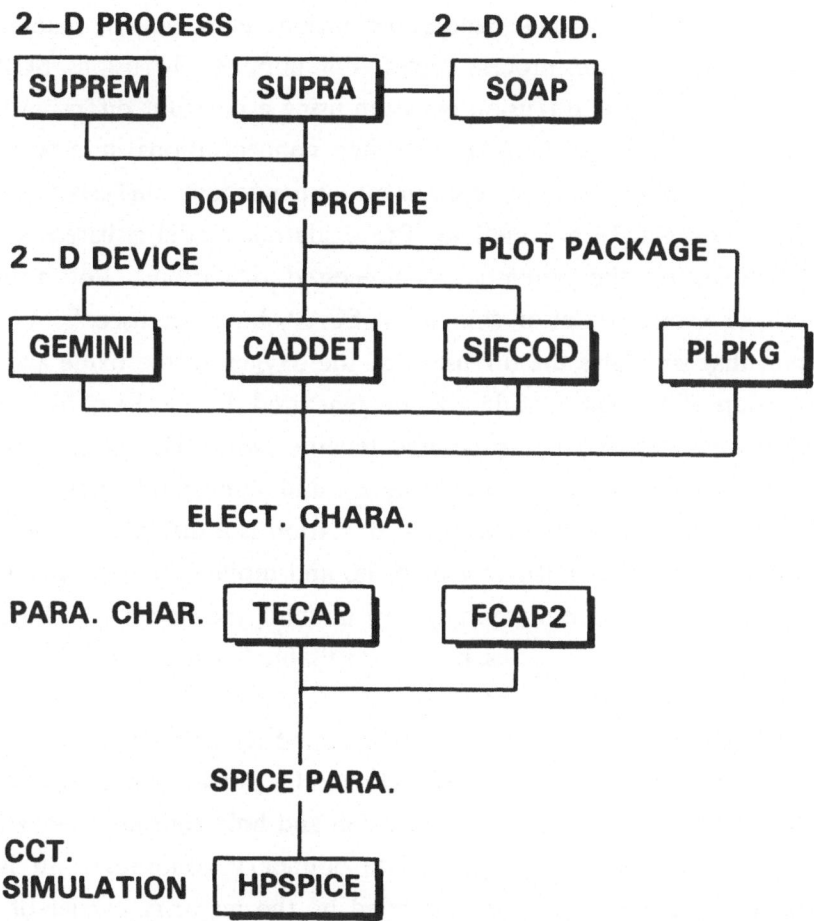

Fig. 1.1 Block diagram of 2-D simulation system.

dimensional, however, SUPREM can be used to generate the doping profile of MOS devices where the shape of 2-D source/drain lateral diffusion is not so critical. In this case, two SUPREM 1-D simulations are needed; one is to specify the doping distribution in the channel and the other is to specify the distribution in the source/drain region. The lateral diffusion is calculated by multiplying the vertical source/drain profile with Gaussian or complementary error functions. SUPRA is needed for more accurate 2-D doping profile in MOS transistors. The 2-D process program, SUPRA, simulates processes based on the device geometry and process schedule and generates the impurity distributions in two dimensions. SUPRA can handle deposition, etch, ion implantation, diffusion and oxidation process cycles. For impurity diffusions, SUPRA analytically solves the diffusion equation using a constant diffusivity for low impurity concentrations. For higher concentrations, it solves the diffusion equation using a concentration-dependent diffusivity in a numerical finite-differnce method. The oxidation model is based on the measurements of the common semi-recessed oxidation. For a more accurate or exotic oxidation simulation, SOAP should be used. SOAP is a program that simulates the diffusion of the oxygen in the oxide and the propagation of the extra oxide volume generated during the oxidation in a rigorous manner. SOAP can be used together with SUPRA to simulate 2-D process including oxidation or as a stand-alone mode just for 2-D oxidation structure. Two-dimensional oxidation is a difficult problem to simulate because of nonplanar geometries and moving boundaries. SOAP uses the boundary-value method. In this method, the nodes are allocated only along the boundary. Thus, it is very suitable for nonplanar geometry and moving-boundary problems.

Electrical device characteristics are predicted by 2-D device simulators based on the impurity distributions predicted by process simulators. In the semiconductor device, the Poisson, electron and hole continuity equations should be solved with the appropriate boundary conditions. In MOS devices, however, most current is carried by the majority carrier of the source/drain. Thus, MOS devices can be simulated by only solving the majority carrier continuity equation. Such an algorithm is used by the full 2-D simulator CADDET. Thus, it can simulate the whole range of the MOS device operation. Due to the simplification, it cannot, however, simulate the phenomena related to the two carrier problem. The simultaneous solution of 2-D Poisson and current continuity equations makes CADDET slow. Simplified analysis should be used whenever possible. CADDET also has limitations in the input structure. For the subthreshold region, where the current is small, it is enough to solve

Poisson's equation with proper guessing of the electron and hole quasi-Fermi level. GEMINI adopts this scheme and speeds up the simulations in the subthreshold region. It has good input and output capability and is fast, but care should be taken to use it within its limits. CADDET and GEMINI have some limitations due to the simplifications. These limitations are acceptable for many MOS device applications. However, they are fatal in some cases, especially in the hot electron related problem, latch-up, or novel devices. To satisfy these applications, a new general-shape 2-D 2-carrier transient device simulator, SIFCOD has been acquired and enhanced. It can simulate the steady-state device characteristics and the switching characteristics together with the lumped circuit elements. It accepts any shape geometry and solves for both carriers so that it can simulate any semiconductor device with almost no limitations.

Based on the device characteristics calculated by these simulators, the electrical parameters can be extracted by TECAP2 [1.16] for use in circuit simulations. TECAP2 measures the device characteristics of MOS transistors and extracts their parameters. Here, the data is taken from the simulations and only the parameter extraction part of TECAP2 is used. In VLSI circuits, accurate determination of interconnect and other capacitance values becomes crucial in the circuit simulation. The value of various capacitance components of MOS circuits can be simulated and determined by FCAP2 [1.17], a two-dimensional, arbitrary-geometry, linear Poisson solver. The circuit performance can be simulated by HP-SPICE [1.18] based on electrical device parameters and capacitances that were obtained from the process schedule and device and circuit layouts using this simulation system.

Each program will be explained in more detail in the subsequent chapters. The process simulators will be discussed in chapter 2 and the device simulators will be in chapter 3. Chapter 4 will deal with parasitics simulation. TECAP2 and SPICE will not be treated because they are not closely related to the topics in this book. The purpose of part I is to introduce these programs to the users as simply as possible so that they can easily understand how these programs work and give them useful examples so that they can start to simulate the practical problems by just modifying the examples without spending too much time to read manuals. The main reason people do not use programs is the bulkiness of the manual. Once a user is familiar how to run the program, he can build up his skill and knowledge with practice and by using the manual. The necessary things for the users to know about each program (its organization, numerical algorithm, and physical model) are briefly explained. After that, practical examples are given to illustrate how to use the program. First, input example files are explained in detail with a brief illustration of the input structure. Common pitfalls are also illustrated to prevent the users from spending too much time without any results.

References

[1.1] M. B. Barron, *"Computer Aided Analysis of IGFET Transistor,"* TR No.5501-1, Stanford Electronics Laboratories, Stanford University, Stanford, CA, Nov 1969.

[1.2] D. Vandorpe and J. Borel, "An Accurate Two-Dimensional Numerical Analysis of the MOS Transistor," *Solid State-Electronics*, 15, 1972, pp 547-557.

[1.3] M. S. Mock, "A Two-Dimensional Mathematical Method of IGFET Transistors," *Solid-State Electronics*, 16, 1973, pp. 601-609.

[1.4] G. D. Hatchel et al, "A Graphical Study of the Current Distribution in Short-Channel IGFETs," *ISSCC Conf. Digest*, Philadelphia, pp. 110-111, 1974.

[1.5] J. J. Barnes and R. J. Lomax, " Two-Dimensional Finite Element Simulation of Semiconductor Devices," *Electron. Lett.*, 10, 8 Aug 1974, pp. 341-343.

[1.6] P. E. Cottrell and E. M. Buturla, " Steady-state Analysis of Field Effect Transistors via the Finite Element Method," *IEDM Tech. Digest*, Dec 1975, pp. 51-54.

[1.7] T. Toyabe, K. Yamaguchi, S. Asai, and M. S. Mock, "A Numerical Model of Avalanche Breakdown in MOSFETs," *IEEE Trans. on Electron Devices*, ED-25, July 1978, pp. 825-831.

[1.8] J. A. Greenfield, S. E. Hansen, and R. W. Dutton, *"Two-Dimensional Analysis for Device Modeling,"* TR No. G201-7, Stanford Electronics Laboratories, Stanford University, Stanford, Calif., 1980.

[1.9] S. Selberherr, W. Fichtner and H. W. Potzl, "Minimos - A Program Package to Facilitate MOS Device Design and Analysis," *Proc. of NASECODE I Conf.* June, 1979, pp. 275-279,

[1.10] M. S. Mock, "A Time-Dependent Numerical Model of the Insulated-Gate Field-Effect Transistor," *Solid-State Electronics* 24, pp. 959-966, 1981.

[1.11] M. R. Pinto et al, " Computer-Aids for Analysis and Scaling of Extrinsic Devices," *IEDM Tech. Digest*, Dec 1984, pp. 288-291.

[1.12] B. E. Deal and A. S. Grove, *J. Appl. Physics.*, *36*, 1965, p. 377

[1.13] D. A. Antoniadis, S. E. Hansen, and R. W. Dutton, *"SUPREM II - A Program for IC Process Modeling and Simulation,"* TR No. 5019-2, Stanford Electronics Laboratories, Stanford University, Stanford, Calif., 1978.

[1.14] H. G. Lee, *"Two-Dimensional Impurity Diffusion Studies: Process Models and Test Structures for Low-Concentration Boron Diffusion,"* TR No. G201-8, Stanford Electronics Laboratories, Stanford University, Stanford, Calif., Aug 1980.

[1.15] D. Chin, M. Kump, and R. W. Dutton, *SUPRA: Stanford University Process Analysis Program,"* Stanford Electronics Laboratories, Stanford University, Stanford, Calif., Oct 1979.

[1.16] E. Khalily, "Transistor Electrical Characterization and Analysis Program," *Hewlett-Packard Journal*, Vol. 32, No. 6, June 1981.

[1.17] Soo-Young Oh, " MOS Device and Process Design Using Computer Simulations," *Hewlett-Packard Journal*, Vol. 33, No. 10, Oct 1982.

[1.18] L. K. Scheffer, R. I. Dowell, and R. M. Apte, "Design and Simulation of VLSI Circuits," *Hewlett-Packard Journal*, Vol. 32, No. 6, June 1981.

Chapter 2
Process Simulation

2.1 Introduction

Silicon integrated circuit (IC) technology has evolved to fabricate multi-million transistors on a single chip. Trial-and-error methodology to optimize such a complex process is no longer desirable because of the enormous cost and turn-around time. From this point of view, computer simulation is a cost-effective alternative, not only supplying a right answer for increasingly tight processing windows, but also serving as a tool to develop future technologies. When coupled with a device analysis program, a process simulator has proven to be a powerful design tool because the process sensitivity to device parameters can be easily extracted by simple changes made to processing conditions in computer inputs. [2.1].

SUPREM (Stanford University PRocess Engineering Models) [2.1] was introduced in 1977 and was the first program capable of simulating most IC fabrication steps. As of 1983, more than 300 copies were distributed world wide and enhanced versions SUPREM II [2.2] and SUPREM III [2.3] are now available. The program accepts a process-runsheet-like input and gives an output containing the impurity distributions in the vertical direction. SUPREM, therefore, can be applied to any regions where impurity distribution changes only in the vertical direction as indicated in a CMOS cross-section (Fig. 2.1). Fabrication of the structure involves several processing steps of ion implantation, oxidation/drive-in and etching/deposition.

The SUPREM program consists of various models based on experimental data as well as physical assumptions which will be discussed in detail in following sections. Some of the models, however, have severe limits in their valid ranges and some are not even fully understood; phosphorous diffusion is an example of this. Parameters associated with these models may also be subjective to individual processing conditions. Users may find occasional discrepancies between measured data and simulation results if they simply use default values in the program. It should be remembered that users need to check whether a model and associated default values in SUPREM are valid for process steps to be simulated. If they are not valid, parameters should be adjusted in order to obtain more accurate results.

With the trend toward shallow junction and lower heat cycles in VLSI technologies, two-dimensional impurity profiles and structures are more crucial to device characteristics. Threshold voltage and parasitic capacitance, for example, are strong functions of lateral diffusion of arsenic in the source/drain and boron in the channel-stop region. Simple extension of SUPREM to two dimensions is not desirable because the degree of equivalent numerical accuracy requires tremendous computing resources. Furthermore, device structures change continuously during such processes as reactive-ion etching and local oxidation of silicon (LOCOS), which impose more difficulties to establishing simulation algorithms. SUPRA (Stanford University PRocess Analysis) [2.4] introduced in 1981 was one of the pioneer works in the two-dimensional process modeling and it can be applied to the areas where impurity distributions and device structures change not only in the vertical direction but in the lateral direction as indicated in Fig. 2.1. It had already been demonstrated that the two-dimensional process simulator could be a powerful design tool when coupled with a device analysis program [2.5]. Nonuniform processing sometimes needs a new concept to explain the physical mechanism. The local oxidation is an example that requires significantly different kinetic equations from the one-dimensional counterpart. SOAP (Stanford Oxidation Analysis

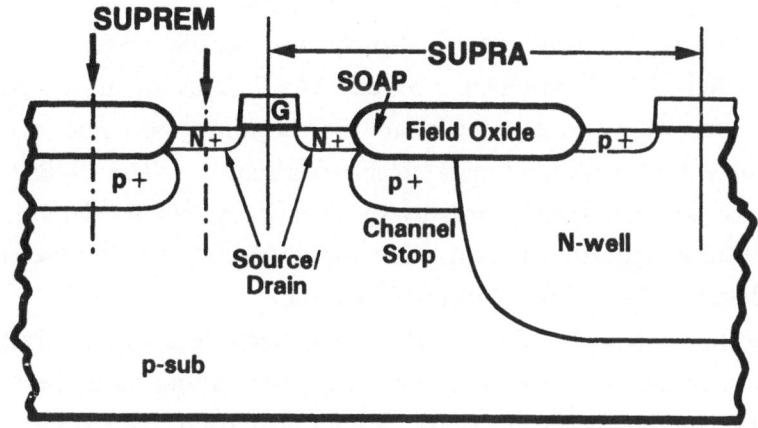

Fig. 2.1 Cross-section of a CMOS device.

Program) [2.6] in 1983 is a program simulating nonuniform oxidation processes.

There may be many other tools [2.7]-[2.9] developed by companies or universities but used only internally or not publicized. In this chapter, we will discuss SUPREM, SUPRA, and SOAP because they are available for public use. The following sections describe physical models and application examples of the programs.

2.2 SUPREM : 1-D Process Simulator

The SUPREM program consists of three main parts: an input scanner supervisor, an output generator, and an actual number-crunching part with various process models. When SUPREM is initiated, it first lists the entire input deck, numbering lines. The input file resembles an actual process runsheet and it consists of free format statements involving key words and numbers. Typically a single processing step can be simulated with an input specification of less than 60 alphanumeric characters. The details of input specification can be referenced in SUPREM II user manual [2.2]. Secondly, it goes through the deck line by line, checking the validity of the syntax. This is done by comparing every command in the input with a key file containing all SUPREM keywords and their related default values. The card order is also checked because certain cards must be preceded. TITLE, GRID and SUBSTRATE cards are always the first three cards that must appear at the beginning because they are responsible for array and parameter initialization. The SUBSTRATE card also initializes model cards by specifying crystalline orientation and the substrate impurity element. The input format will be discussed in more detail in the following example section.

When impurities are introduced in subsequent steps, proper arrays and physical parameters are selected and calculations are performed, depending upon processing steps. Ion implantation, oxidation, etching, deposition, impurity predeposition, and epitaxial growth are treated in SUPREM. The model cards allow users to modify internal program coefficients associated with the process steps as well as impurity elements (B, P, Sb and As). The output of the program, available at the end of each step, consists of the one-dimensional profiles of all the dopants present in the silicon and silicon-dioxide materials. The profiles may be displayed in various formats on a line printer or a high resolution plotter.

Process Models

(a) Ion Implantation

The simplest description of an implanted profile in silicon dioxide is a symmetric Gaussian curve with the first two moments (the projected range R_p, and the standard deviation, σ_p) calculated from LSS theory [2.10]. The actual distributions of many ions are better fitted by two half-Gaussian profiles, each with different deviation, σ_1 and σ_2, joined together at a modal range R_m. The joint half-Gaussian expression is used in SUPREM for As, P, and Sb as follows,

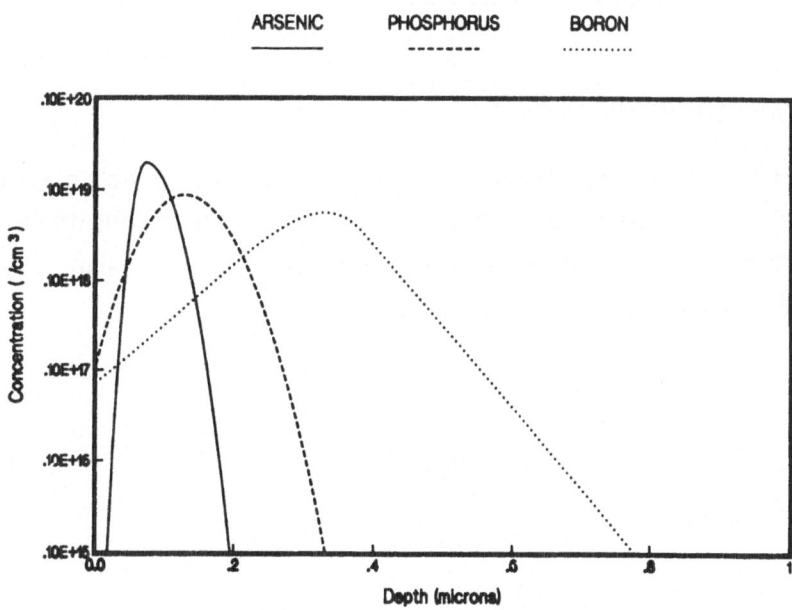

ARSENIC   PHOSPHORUS   BORON

Fig. 2.2 Profiles of B, As and P with same energy and dose.

$$C_1(y) = C_p \exp[-(y - R_m)^2/2\sigma_1^2] \qquad 0 \le y \le R_m \quad (2.1a)$$

$$C_2(y) = C_p \exp[-(y - R_m)^2/2\sigma_2^2] \qquad R_m < y \le \infty \quad (2.1b)$$

where C_p is the peak concentration. SUPREM calculates the three parameters R_m, σ_1, and σ_2 using a look-up table containing R_p, σ_p and the third moment as a function of energy.

An implanted boron profile, however, shows a long tail because of channeling. It is described by an empirically modified Pearson IV with the fourth moment and an exponential tail. Pearson distribution f is a generalized function to fit any practical distributions as defined below

$$\frac{1}{f}\frac{df}{dy} = \frac{a - y}{b_0 + b_1 y + b_2 y^2} \qquad (2.2)$$

where coefficients a, b_0, b_1, and b_2 can be arbitrarily chosen. Fig. 2.2 illustrates different characteristics of arsenic, phosphorus and boron impurities implanted with the same energy and dose. Since boron is the lightest atom among them, the implantation peak is located at the deepest position.

b) Thermal Oxidation

Silicon dioxidation (oxidation) is a thermal process in which oxidizing species diffuse through an oxide layer and react with silicon atoms. A 125 percent volume expansion is accompanied. Three oxidant fluxes involved in the oxidation process are

$$F_1 = h(C^* - C_o) \qquad (2.3a)$$

$$F_2 = D_{eff} \frac{C_o - C_i}{X_o} \qquad (2.3b)$$

$$F_3 = kC_i \qquad (2.3c)$$

where
  F_1 = transport flux from gas ambient to the oxide surface
  F_2 = diffusion flux inside the oxide layer
  F_3 = reaction flux at the silicon/oxide interface
  C^* = equilibrium concentration in the oxide
  C_o = concentration at the oxide surface
  C_i = concentration at the silicon/oxide interface
  h = gas transport coefficient
  D_{eff} = effective diffusion coefficient
  X_o = oxide thickness
  k = surface reaction coefficient
The linear-parabolic oxide-growth model [2.11] in SUPREM assumes steady-state oxidant diffusion that the three fluxes are equal as

$$F_1 = F_2 = F_3 = F \qquad (2.4)$$

The oxide growth rate is directly proportional to the flux as

$$\frac{dX_o}{dt} = \frac{F}{N_1} = \frac{kC_i/N_1}{1 + k/h + kX_o/D_{eff}} \qquad (2.5)$$

where N_1 is the number of oxidant molecules incorporated in a unit volume of the oxide layer. When integrated, Eq. (2.5) leads to the well-known linear-parabolic growth relationship but only if an initial oxide X_i is specified prior to the oxidation step under consideration

$$\frac{X_o^2 - X_i^2}{B} + \frac{X_o - X_i}{B/A} = t \qquad (2.6)$$

where we call B the parabolic rate constant and B/A the linear rate constant. Under relatively low dopant concentration conditions, B and B/A depend only on silicon crystal orientation, oxidizing ambient and temperature. Behaviors of the two rate constants stored in SUPREM as default values are shown in Fig. 2.3. Users need to compare their own values because the rate constants are also sensitive to such oxidation conditions as partial pressure of oxygen, HCl content and temperature ramping.

Under high surface impurity concentration, more silicon vacancies are created, and the rate constants are, consequently, enhanced [2.12]. Since the impurity concentration at the silicon/oxide interface changes due to diffusion and segregation during oxidation, enhanced values are calculated at each time step in SUPREM. It is known that the growth rate is enhanced as much as a factor of 10 for oxide thinner than 20 nm in a dry oxygen condition. This fast oxidation phenomenon has attracted great deal of attention recently because the gate oxide in the present VLSI process is grown within the thin oxide regime. SUPREM III added an empirical factor to the linear-parabolic model as

$$\frac{dX_o}{dt} = \frac{B}{X_o + A} + K \exp(-\frac{X_o}{L}) \qquad (2.7)$$

where the decay length L is approximately independent of temperature (~ 7 nm) and K is a singly activated function of temperature with an activation energy of 2.35 eV for <111> and <100> orientations and 1.8 eV for <100> silicon [2.13].

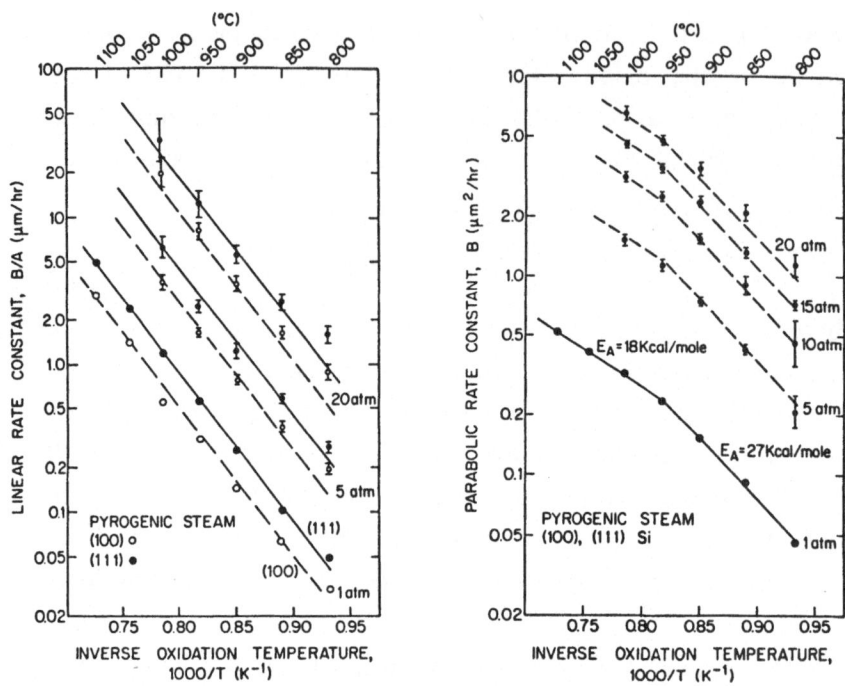

Fig. 2.3 Temperature dependence of oxide growth rates.

c) Impurity Redistribution

Impurity diffusion in silicon during high-temperature processing steps can be described by the complete continuity equation

$$\frac{\partial C}{\partial t} = \frac{\partial}{\partial x}(D\frac{\partial C}{\partial x}) \pm \frac{-q}{kT}\frac{\partial}{\partial x}(DC_i\frac{\partial \phi}{\partial x}) \qquad (2.8)$$

where D is the diffusivity, and C and C_i are the total and electrically charged impurity concentrations, respectively. The potential ϕ is

$$\phi = \frac{kT}{q}\ln\frac{n}{n_i} \qquad (2.9)$$

where n and n_i are the electron and intrinsic carrier concentrations, respectively. The first term in the continuity equation represents classical concentration gradient-driven diffusion, including non-constant diffusivity. The second term incorporates the electrostatic field-driven flux.

SUPREM utilizes models based on vacancy diffusion mechanisms under non-oxidizing conditions. The intrinsic diffusivity of an ionized impurity species is the sum of the diffusivities resulting from neutral vacancies and ionized vacancies with an opposite charge. There are four charged states for vacancies: doubly negative (=), singly negative (-), neutral (x) and positive (+). Thus, the effective diffusivity under non-oxidizing conditions

$$D_N = D_i^{\times} + D_i^{-}[V^-] + D_i^{=}[V^=] + D_i^{+}[V^+] \qquad (2.10)$$

A boron atom as an acceptor is negatively charged in the silicon lattice and it diffuses primarily with V^+ and $V^{\times}$ vacancies. Thus the boron diffusivity is

$$D_N(B) = D_i^{\times} + D_i^{+}(\frac{p}{n_i})$$
$$= [0.037 + 0.72(\frac{p}{n_i})] \exp(-\frac{3.46eV}{kT}) \quad cm^2/\sec \qquad (2.11)$$

The diffusion coefficients for boron and other impurities are given as default values in SUPREM III [2.3]. Arsenic as a donor appears to diffuse with $V^{\times}$ and V^- vacancies and its diffusivity is

$$D_N(As) = D_i^{\times} + D_i^{-}(\frac{n}{n_i}) = 0.066 \exp(-\frac{3.44eV}{kT})$$
$$+12.0 (\frac{n}{n_i}) \exp(-\frac{4.05eV}{kT}) \quad cm^2/\sec \qquad (2.12)$$

It is known that some impurities form immobile complexes. The arsenic clustering effect [2.14] that an As ion forms a pair with charged vacancies when doping level approaches the solid solubility is also considered in SUPREM. For phosphorus diffusion, SUPREM adopts Fair and Tsai's model [2.15] that predicts with reasonable accuracy the phosphorous kink formation as well as the base push effect, commonly observed during heavy emitter diffusions in bipolar technology. The model basically explains these 'anomalous' effects by enhancement of vacancy concentration in the silicon by the dissociation of the phosphorus-doubly ionized vacancy pairs that flow from the surface into the silicon bulk. Near the surface, P, as a donor, diffuses with V^x, V^- and $V^=$ vacancies

$$D_N(P) = D_i^x + D_i^- (\frac{n}{n_i}) + D_i^= (\frac{n}{n_i})^2$$

$$= 3.85 \exp(- \frac{3.66eV}{kT}) + 4.44 (\frac{n}{n_i}) \exp(- \frac{4.0eV}{kT}) \qquad (2.13)$$

$$+ 44.2 (\frac{n}{n_i})^2 \exp(- \frac{4.37eV}{kT}) \quad cm^2/ sec$$

The total phosphorus and electron concentrations are

$$C_T(P) = n + 5.33 \times 10^{-43} n^3 \exp(\frac{0.4 \, eV}{kT}) \qquad (2.14)$$

The bandgap narrowing effect due to lattice-misfit strain is also introduced in SUPREM III, which causes the diffusivity to decrease in heavily doped regions (>5E20 cm^{-3}). At the tail region of P profile, the diffusivity is enhanced relative to the intrinsic value due to super-saturation of the silicon lattice by vacancies from the $P^+ V^=$ dissociation. The diffusivity at the tail is given by

$$D_{tail} (P) = D_i^x + D_i^- \frac{n_s^3}{n_e^2 \, n_i} \exp(\frac{3\Delta E_g}{kT})$$

$$[1 + \exp(\frac{0.3eV}{kT})] \exp(- \frac{X - X_e}{L_v}) \qquad (2.15)$$

where n_s is the surface electron concentration, n_e is the electron concentration at which Fermi level drops below 0.11 eV from the conduction band edge, X_e is the depth at which n_e is reached, and L_v is the diffusion length of the vacancies into the substrate (~25 μm). Figure 2.4 shows redistribution profiles of three implanted dopants in Fig. 2.2, where the impurities were diffused separately at the same temperature and time (1000°C and 30 min).

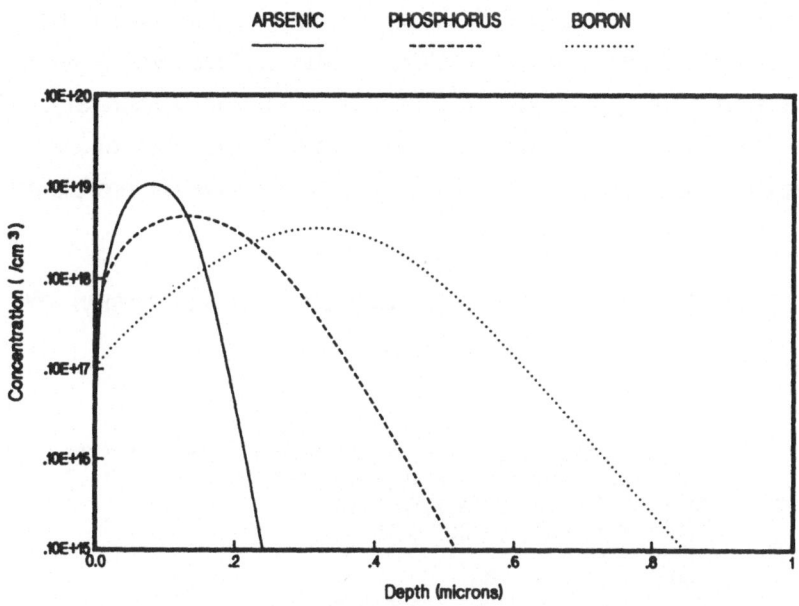

Fig. 2.4 Redistributed profiles of As, P, and B.

Application Examples

Two simulation examples associated with n-channel MOSFET fabrication are presented in this section; first the channel region and second the source/drain region in Fig. 2.1. An input file for the first example is shown in Fig. 2.5, in which the first card TITLE describes the content of following simulations. The material properties of the substrate are given in the SUBSTRATE card which includes the parameters of element, concentration and orientation. ELEMENT specifies the impurity type that can be As, P, B or Sb, or simply n-type (-) or p-type (+). The n-type or p-type specification assumes that the substrate impurity will not be included in the redistribution calculation

during the subsequent processes. In this example, silicon substrate is <100>
crystal orientation and doped by boron whose concentration is
6E14 cm^{-3}.

The following GRID card describes how to set up a initial grid set for
numerical solution. In SUPREM, the grid is divided into two uniformly
spaced regions. One is the area close to the surface where impurity
concentrations rapidly change and thus high resolution is necessary. The
other is the region with low resolution beneath the first one. DYSI
specifies the grid spacing in the high resolution and DPTH represents its

```
          *** STANFORD UNIVERSITY PROCESS ENGINEERING MODELS PROGRAM ***

                            *** VERSION 0-05 ***


    1....     TITLE CHANNEL PROFILE SIMULATION OF N-CHANNEL MOSFET
    2....     SUBSTRATE ORNT=100,ELEM=B,CONC=6E14
    3....     GRID DYSI=0.005,DPTH=1,YMAX=10
    4....     MODEL NAME=DRY1 , PRES=0.05
    5....     PRINT HEAD=Y
    6....     COMMENT DEPOSIT SRO
    7....     STEP TYPE=DEPO,TIME=1,GRTE=0.04
    8....     COMMENT IMPLANT THROUGH SRO
    9....     STEP TYPE=IMPL,ELEM=B,DOSE=5E11,AKEV=30
   10....     STEP TYPE=IMPL,ELEM=B,DOSE=4E11,AKEV=70
   11....     COMMENT ETCH SRO
   12....     STEP TYPE=ETCH,TEMP=25
   13....     COMMENT DRIVE IN
   14....     STEP TYPE=OXID,TEMP=1000,TIME=15
   15....     COMMENT DRIVE IN AND GROW GATE OXIDE
   16....     STEP TYPE=OXID,TEMP=1000,TIME=34,TRTE=-3
   17....     STEP TYPE=OXID,TEMP=900,TIME=5,MODL=DRYO
   18....     STEP TYPE=OXID,TEMP=900,TIME=9,MODL=STMO
   19....     STEP TYPE=OXID,TEMP=900,TIME=25,MODL=DRY1
   20....     COMMENT POLY DOPING
   21....     STEP TYPE=OXID,TEMP=950,TIME=18
   22....     COMMENT S/D REOXIDATION + DENSIFICATION
   23....     STEP TYPE=OXID,TEMP=900,TIME=130
   24....     PLOT TOTL=Y,WIND=1.5
   25....     COMMENT LPCVD DEPOSITION
   26....     STEP TYPE=OXID,TEMP=905,TIME=60
   27....     SAVE FILE=DL026,TYPE=B
   28....     END
    1
```

Fig. 2.5 SUPREM input for NMOS channel stop.

depth. YMAX is the maximum calculation thickness of the silicon, and the thickness of the low-resolution region, therefore, is (YMAX-DPTH). All input dimensions for SUPREM are in micrometers. The grid space of the low-resolution region is twice of that of the high-resolution region. If the number of grid points necessary with the space exceeds the limit (400), the program keeps the number by expanding the grid space.

The MODEL card allows a user to modify default parameter values for special process steps. The model can be named by NAME so that it can be cited without repeating the model card. In this example, the partial pressure of oxygen is modified by PRES = 0.05 to simulate dry oxidation with 5% oxygen and 95% nitrogen ambient. The next card, PRINT, sets logical flags which will print output results after each step. Next, HEAD controls the printing of heading information. The COMMENT card is used only for documentation purpose. The STEP card in the 7th line is to deposit a stress-relief oxide layer where TYPE, TIME, and GRTE indicate the process type, deposition time, and oxide growth rate (μm/min.), respectively. Double channel implantation through the oxide layer is performed by two implant STEP cards. Impurity type is given by ELEMENT, total dose (atoms/cm^2) by DOSE and the energy in KeV by KEV. The next STEP card in line 12 is to etch the oxide layer.

The STEP cards of oxidation from line 14 to 26 describe high temperature thermal cycles of drive-in, gate oxide growth, polysilicon doping, source/drain reoxidation, densification, and LPCVD deposition related to an actual NMOS fabrication process. TRTE represents the temperature ramping rate in degree/min., and MODL specifies the oxidation model used in the following process step. When MODL is DRY0 or STM0, default values for dry or steam oxidation are used. The PLOT card in line 24 sets the flag to make plots of impurity profiles at following steps. In this example, the total impurity concentration is plotted from the oxide surface to 1.5 μm inside the substrate. The range of y-axis is specified by the default value (7 decades starting from 1E14 cm^{-3}). Finally, boron distribution is saved to the file of DL026 in line 26. The END card signals the end of the input file. The SUPREM calculation results of the first example are shown in Fig. 2.6 and 2.7. The first part of Fig. 2.6 is the heading, followed by a summary of each step. Device parameters of junction depth and sheet resistance are also listed. Concentrations of total and active impurities are integrated and printed.

After the oxidation step in line 26, the total impurity concentration is plotted from the oxide surface to the point 1.5 μm inside the silicon as specified by the PLOT card. Care should be taken that the scales in the oxide and the silicon regions are different. Fig. 2.6 shows the boron distribution after all the thermal cycles.

The second example is to simulate the As profile in the source/drain region, which is shown in Fig. 2.8. This can be done by adding arsenic implantation. The dose is $6E15$ cm^{-2} and the energy is 80 KeV. The final impurity profile that went through all the same heat cycles is plotted in Fig. 2.9, where the junction depth is about 0.2 μm.

```
CHANNEL PROFILE SIMULATION OF N-CHANNEL MOSFET
IMPLANT THROUGH SRO
STEP # 3


ION IMPLANT (PEARSON TYPE IV DISTRIBUTION)
IMPLANTED IMPURITY = BORON
IMPLANTED DOSE      = 4.000000E+11
IMPLANT ENERGY      =   70.0000
RANGE               =    .219190
STANDARD DEVIATION = 6.760000E-02
PEAK CONCENTRATION = 2.526850E+16

SURFACE CONCENTRATION = 1.561468E+16 ATOMS/CM^3

    JUNCTION DEPTH      !     SHEET RESISTANCE
-----------------------!-----------------------------
                       !   9559.77     OHMS/SQUARE


NET ACTIVE CONCENTRATION

OXIDE   CHARGE = 2.799563E+10    IS    1.87    % OF TOTAL
SILICON CHARGE = 1.472005E+12    IS    98.1    % OF TOTAL
TOTAL   CHARGE = 1.500001E+12    IS    136.    % OF INITIAL
INITIAL CHARGE = 1.100000E+12


CHEMICAL CONCENTRATION OF BORON

OXIDE   CHARGE = 2.799563E+10    IS    1.87    % OF TOTAL
SILICON CHARGE = 1.472005E+12    IS    98.1    % OF TOTAL
TOTAL   CHARGE = 1.500001E+12    IS    136.    % OF INITIAL
INITIAL CHARGE = 1.100000E+12
```

Fig. 2.6 SUPREM output heading of the input file in Fig. 2.5

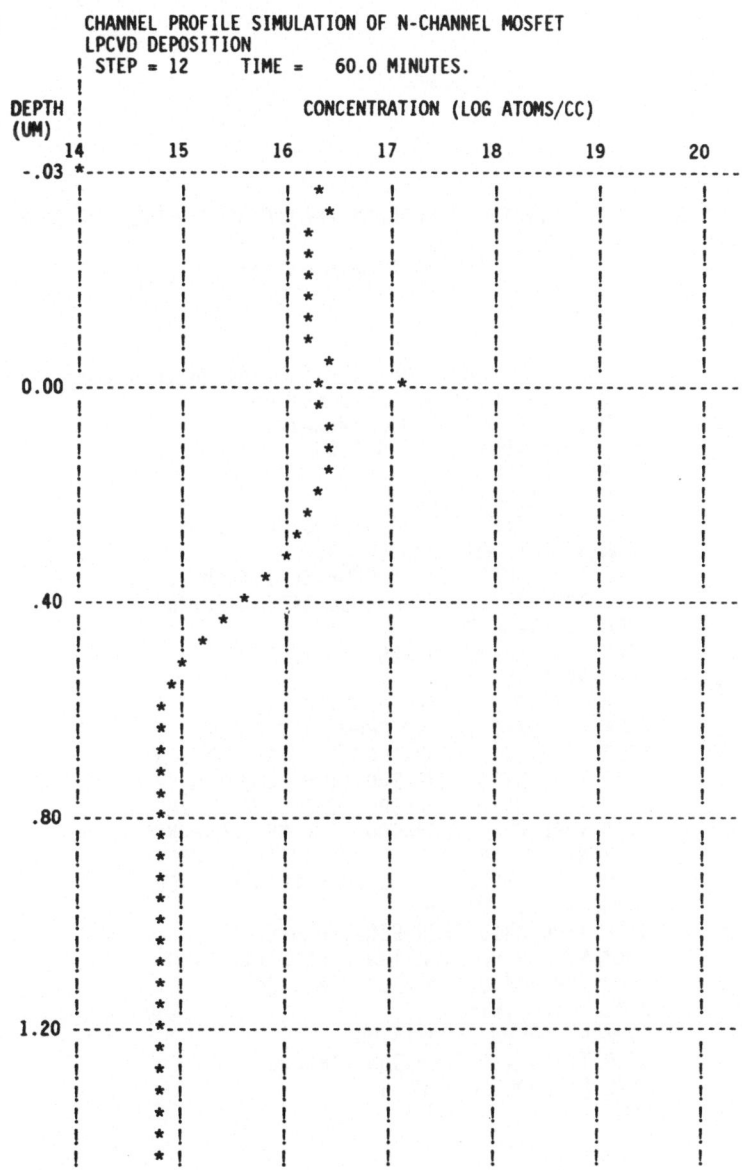

Fig. 2.7 SUPREM output of boron distribution

```
        *** STANFORD UNIVERSITY PROCESS ENGINEERING MODELS PROGRAM ***

                         *** VERSION 0-05 ***


  1....    TITLE SOURCE/DRAIN PROFILE SIMULATION OF N-CHANNEL MOSFET
  2....    SUBSTRATE ORNT=100,ELEM=B,CONC=6E14
  3....    GRID DYSI=0.005,DPTH=1,YMAX=10
  4....    MODEL NAME=DRY1 , PRES=0.05
  5....    PRINT HEAD=Y
  6....    COMMENT DEPOSIT SRO
  7....    STEP TYPE=DEPO,TIME=1,GRTE=0.04
  8....    COMMENT IMPLANT THROUGH SRO
  9....    STEP TYPE=IMPL,ELEM=B,DOSE=5E11,AKEV=30
 10....    STEP TYPE=IMPL,ELEM=B,DOSE=4E11,AKEV=70
 11....    COMMENT ETCH SRO
 12....    STEP TYPE=ETCH,TEMP=25
 13....    COMMENT DRIVE IN
 14....    STEP TYPE=OXID,TEMP=1000,TIME=15
 15....    COMMENT DRIVE IN AND GROW GATE OXIDE
 16....    STEP TYPE=OXID,TEMP=1000,TIME=34,TRTE=-3
 17....    STEP TYPE=OXID,TEMP=900,TIME=5,MODL=DRYO
 18....    STEP TYPE=OXID,TEMP=900,TIME=9,MODL=STMO
 19....    STEP TYPE=OXID,TEMP=900,TIME=25,MODL=DRY1
 20....    COMMENT S/D IMPLANT THROUGH GATE OXIDE
 21....    STEP TYPE=IMPL,ELEM=AS,DOSE=6E15,AKEV=80
 22....    COMMENT POLY DOPING
 23....    STEP TYPE=OXID,TEMP=950,TIME=18
 24....    COMMENT S/D REOXIDATION + DENSIFICATION
 25....    STEP TYPE=OXID,TEMP=900,TIME=130
 26....    PLOT TOTL=Y,WIND=1.5
 27....    COMMENT LPCVD DEPOSITION
 28....    STEP TYPE=OXID,TEMP=905,TIME=60
 29....    SAVE FILE=DLO26,TYPE=B
 30....    END
```

Fig. 2.8 SUPREM input for the source/drain profile

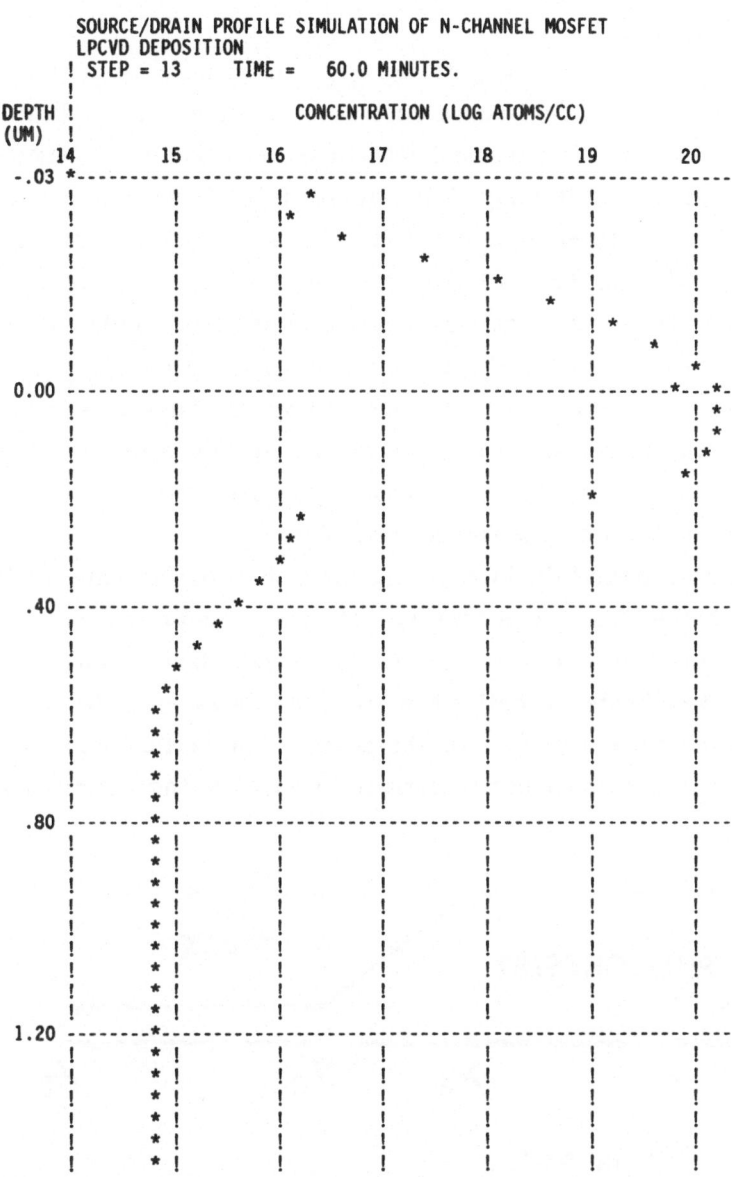

```
SOURCE/DRAIN PROFILE SIMULATION OF N-CHANNEL MOSFET
LPCVD DEPOSITION
! STEP = 13      TIME =   60.0 MINUTES.
!
DEPTH !                 CONCENTRATION (LOG ATOMS/CC)
(UM)  !
      14       15        16        17        18        19        20
 -.03 *-------------------------------------------------------------
      !         !         !  *      !         !         !         !
      !         !         !*        !         !         !         !
      !         !         !      *  !         !         !         !
      !         !         !         !   *     !         !         !
      !         !         !         !   !*    !         !         !
      !         !         !         !         !  *      !         !
      !         !         !         !         !       ! *        !
      !         !         !         !         !         !  *   ! !
      !         !         !         !         !         !      *!
 0.00 ------------------------------------------------------------*---*-
      !         !         !         !         !         !       !  *
      !         !         !         !         !         !       !  *
      !         !         !         !         !         !       !*  *
      !         !         !         !         !         !  *    *!
      !         !         !  !*     !         !         !       !
      !         !         !*        !         !         !       !
      !         !         !  *      !         !         !       !
      !         !        * !        !         !         !       !
  .40 ------------------*-----------------------------------------------
      !         !   !  *    !         !         !         !       !
      !         !  *        !         !         !         !       !
      !         !  *        !         !         !         !       !
      !         *!          !         !         !         !       !
      !         *!          !         !         !         !       !
      !         *!          !         !         !         !       !
      !         *!          !         !         !         !       !
      !         *!          !         !         !         !       !
  .80 ----------*------------------------------------------------------
      !         *!          !         !         !         !       !
      !         *!          !         !         !         !       !
      !         *!          !         !         !         !       !
      !         *!          !         !         !         !       !
      !         *!          !         !         !         !       !
      !         *!          !         !         !         !       !
      !         *!          !         !         !         !       !
      !         *!          !         !         !         !       !
 1.20 ----------*------------------------------------------------------
      !         * !         !         !         !         !       !
      !         * !         !         !         !         !       !
      !         * !         !         !         !         !       !
      !         * !         !         !         !         !       !
      !         * !         !         !         !         !       !
```

Fig. 2.9 Total impurity profile calculated by SUPREM

2.3 SUPRA : 2-D Process Simulator

SUPRA simulates the incorporation and redistribution of impurities in a two-dimensional (2-D) cross-section of a device as indicated in Figure 2.1. Such two-dimensional structures are created by resembling actual lithographic patterning. A photoresist layer deposition in Fig. 2.10, for example, is removed by an ETCH card. Four parameters (START, CORNER, END and THICK) gives enough freedom to create arbitrary etch profiles. However, re-entrant angles resulting from undercutting are not allowed because such a shape is difficult to express numerically with an array that represents a single value of a layer thickness at a horizontal position. Impurity concentrations are calculated only within the thermal oxide and substrate regions. The mask layers are used only as barriers against ion implantation and surface diffusion.

Input commands and the internal organization of SUPRA are similar to those of SUPREM. Specification of the two-dimensional device structures, however, requires the use of a coordinate system. The horizontal coordinate is defined as x and corresponds to variation parallel to the bottom surface of the device. The vertical coordinate is defined as y and corresponds to variation normal to the bottom surface

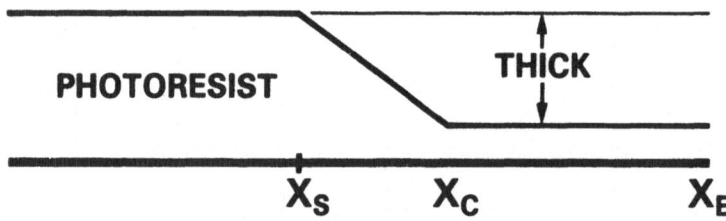

X_S : start
X_C : corner
X_E : end

Fig. 2.10 Structure definition in SUPRA. START, CORNER, and END represent the starting, corner, and ending positions. THICK is the thickness of the layer to be etched.

of the device. The origin of coordinates is defined as the leftmost point of the original semiconductor/oxide interface, with positive x to the right and positive y downward into the substrate.

Finite-difference grids in two dimensions are easily generated by using X.GRID and Y.GRID cards for the x and y directions. Nonuniform grids are automatically generated, depending upon the number of nodes and spaces specified in each X.GRID or Y.GRID card. This allows fast computation as well as high resolution for regions where concentrations change rapidly. More details on the grid generation will be discussed in a following example section.

Two-Dimensional Process Models

Process models of the SUPRA program consist of two main parts. One is an analytical part containing closed forms of analytical equations, from which solutions are readily obtained at any position without numerical iterations. The other is a numerical part in charge of specific calculations for solving the nonlinear diffusion equation. Analytical solutions are used to model changes in the overall device structure (i.e. deposition/etch and oxide growth), ion implantation, and boron diffusion. This analytic technique is fast in computation and flexible in grid allocation. It is, however, limited to low-concentration diffusion. Inert drive-in involving high impurity concentration of arsenic-implanted source/drain regions is treated by the numerical solution.

a) Ion Implantation

The spatial distribution of implanted atoms is assumed to be a Gaussian function in the vertical and lateral direction when implanted through an infinitesimal mask window (actually a point source). When this Gaussian function is integrated for a finite window with a uniform thickness as shown in Fig. 2.11 from x_k to x_{k+1}, the profile is given below as

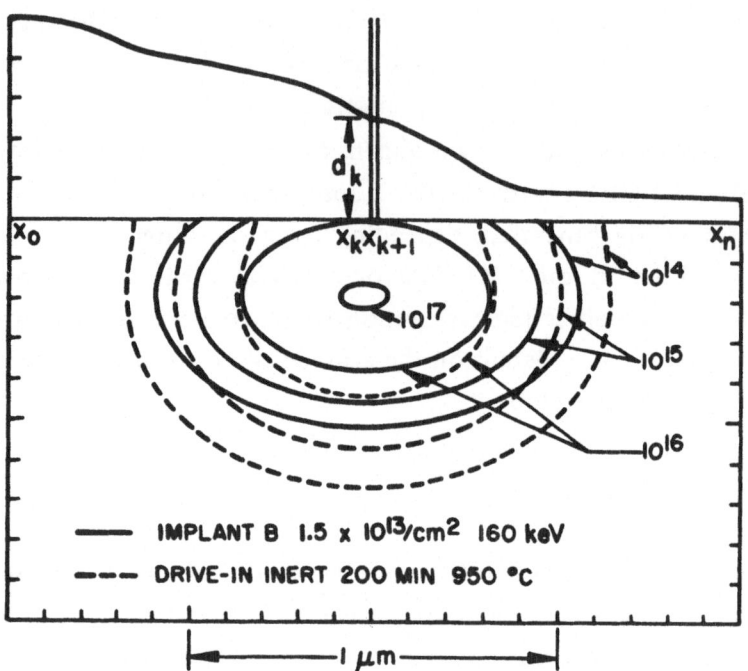

Fig. 2.11 Impurity distributions implanted through a
small mask window after implantation (solid
lines) and after inert drive-in (broken lines).

$$I_k(x,y) = \frac{I_{max}}{2} \exp[\frac{-(y-R_p-d_k)^2}{2\Delta R_p^2}]$$

$$\cdot\{-\text{erfc}(\frac{x-x_k}{\sqrt{2}\,\Delta x}) + \text{erfc}(\frac{x-x_{k+1}}{\sqrt{2}\,\Delta x})\}$$

$$(2.16)$$

where I_{max} is the peak concentration and $\text{erfc}(x)$ is the complementary
error function of x. The parameters R_p, ΔR_p and Δx are the projected
range, and vertical and lateral standard deviations respectively. It should
be noted that the profile implanted through the window with a finite
width becomes a Gaussian distribution multiplied by a
complementary-error function in the lateral direction. The impurity
profile, when implanted through an arbitrary mask as shown in Fig. 2.11,
can be expressed by a summation of the profiles through all mask

segments as

$$I(x,y) = \sum_{k=1}^{n} I_k(x,y) \tag{2.17}$$

where k represents the kth mask segment.

b) Low Concentration Inert Drive-in

If the impurity concentration is lower than the intrinsic carrier concentration n_i, as in the boron channel-stop region, the diffusivity can be assumed constant. The diffusion Eq. (2.8) becomes linear when the field term is ignored. The same superposition technique used for implantation can be applied to this inert drive-in condition. The inert drive-in solution for the profile described in Eq. (2.16) is given by

$$N_i(x,y,t) = N_{ix}(x,t)N_{iy}(y,t) \tag{2.18a}$$

$$N_{ix}(x,t) = -\frac{1}{2}\text{erfc}(\frac{x - x_k}{\sqrt{2\Delta x^2 + 4Dt}})$$
$$+ \frac{1}{2}\text{erfc}(\frac{x - x_{k+1}}{\sqrt{2\Delta x^2 + 4Dt}}) \tag{2.18b}$$

$$N_{iy}(y,t) = \frac{I_{max}\Delta R_p}{2\sqrt{2Dt + \Delta R_p^2}}\{\Omega(y,t) + \Omega(-y,t)\} \tag{2.18c}$$

$$\Omega(y,t) = \exp\{-\frac{(y - R_p - d_k^2)}{4Dt + 2\Delta R_p^2}\}$$
$$[2 - \text{erfc}\{\frac{y\Delta R_p^2 + 2(R_p - d_k)Dt}{\Delta R_p\sqrt{2Dt(4Dt + 2\Delta R_p^2)}}\}] \tag{2.18d}$$

where the subscript i refers to inert ambient conditions. D is the impurity diffusivity, t is the time of the diffusion step, d_k is the effective silicon thickness of the layers above the substrate through which the ion implantation is performed. The profile resulting from an inert drive-in of the implant through one mask segment is shown in Fig. 2.11 (broken lines). Once again, the complete drive-in profile is constructed by superposing the solution for each mask segment.

c) Moving Boundary Diffusion

The one-dimensional profile following oxidation is approximated by adding a correction factor to the inert drive-in concentration [2.17] as

$$N_o(y,t) = N_i(y,t) + N_c(y,t) \tag{2.19}$$

where the correction factor $N_c(y,t)$ is expressed by a term involving a complementary error function. In this approach the correction function basically subtracts an appropriate fraction of dopant to account for segregation and out-diffusion into the oxide. A boundary condition which properly determines the correction term has been found, and extensive analytical and numerical results have been presented elsewhere [2.18]. For low concentration diffusion, this analytic method gives agreement typically better than 10% with numerical calculations, and it has a computation-time advantage of more than a factor of ten [2.19].

A quasi two-dimensional profile has been obtained by assuming that the oxide layer grows only in the vertical direction during a semi-recessed oxidation [2.18]. However, the effect of lateral oxide growth may not be negligible in a case such as fully-recessed oxidation. A more generalized boundary condition is given below as

$$D\nabla N_o(x,y,t)\big|_{x_f, y_f} = A_o N_o(x_f, y_f, t) g(x_f, y_f, t) \mathbf{n}(x_f, y_f, t) \tag{2.20a}$$

$$A_o = \frac{1}{m} - \alpha \tag{2.20b}$$

where the coefficient m is the equilibrium segregation factor and α is the volumetric ratio of silicon consumed in forming one unit of oxide (0.44). The quantity $g(x_f,y_f,t)$ is the oxide growth rate, $\mathbf{n}(x_f,y_f)$ is a unit vector normal to the SiO_2 interface, and (x_f,y_f) is the final interface point closest to (x,y) as indicated in Fig. 2.12. Provided that $N_o(x,y,t)$ can be separated in the x and y variables, the above boundary condition is also separable in the vertical and lateral directions. The resulting 2-D profile is given by

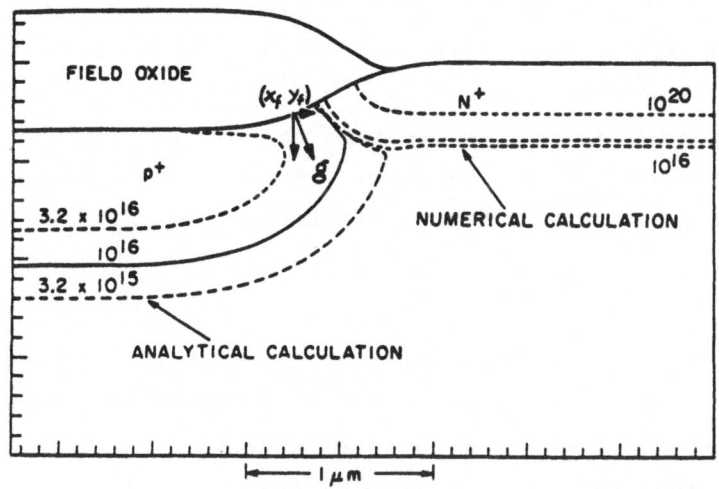

Fig. 2.12 Cross-section of NMOS FET near the source and the channel-step region simulated by SUPRA.

$$N_o(x,y,t) = N_{ox}(x,t)N_{oy}(y,t) \qquad (2.21a)$$

$$N_{ox}(x,t) = N_{ix}(x,t)\{1 + A_x\mathrm{erfc}(\frac{x-x_f}{2\sqrt{Dt_{ox}}})\} \qquad (2.21b)$$

$$N_{oy}(y,t) = N_{iy}(y,t)\{1 + A_y\mathrm{erfc}(\frac{y-y_f}{2\sqrt{Dt_{ox}}})\} \qquad (2.21c)$$

where the subscripts o and i indicate oxidizing and inert conditions, respectively. The Dt terms implicit in N_{ix} and N_{iy} represent the diffusivity-time product for all high temperature steps including inert drive-in. The correction terms involve only the moving boundary and segregation effects, hence Dt_{ox} represents the diffusivity-time product for oxidation steps only.

The coefficient A_x and A_y are obtained from Eq. (2.21) and the boundary condition (2.20) as

$$A_z = \frac{A_o g_z N_{iz}(z_f, t) - D\frac{\partial N_{iz}(z,t)}{\partial z}|_{z=z_f}}{(A_o g_z + D/\sqrt{\pi Dt_{ox}})N_{iz}(z_f, t) - D\frac{\partial N_{iz}(z,t)}{\partial z}|_{z=z_f}} \qquad (2.22)$$

where z is either the x or y variable. Again g_z is the oxide growth rate in either the vertical or lateral direction. The boron profile obtained by this analytical technique at the channel-stop region is shown in Fig. 2.12. The two-dimensional oxide shape resulted from LOCOS is approximated by empirically determined functions. More rigorous treatment of non-uniform oxidation is the topic of the next section.

d) High-Concentration Diffusion

If the impurity concentration is sufficiently large compared to the intrinsic electron concentration at the diffusion temperature, then diffusivity will no longer be constant throughout the simulation space. The diffusion equation then becomes nonlinear and must be solved numerically. Also, there may be several impurities present simultaneously which can interact by way of electric field or Fermi level effects, whereby impurity clustering can become important.

The diffusion equation for each impurity species is obtained by extending Eq. (2.8) to two dimensions as

$$\frac{\partial C}{\partial t} = - \nabla \cdot \mathbf{J} \qquad (2.23a)$$

$$\mathbf{J} = - D\nabla N \pm \frac{-q}{kT}(DN\nabla\phi) \qquad (2.23b)$$

$$\phi = \frac{kT}{q} \ln(\frac{n}{n_i}) \qquad (2.23c)$$

where the basic models on the diffusivity D and the total/active concentrations are the same as SUPREM discussed in section 2.2. The potential ϕ can be eliminated by using Eq. (2.9) in which the electron concentration can be expressed by the algebraic sum of all active impurity concentrations (U) as

$$n = \frac{U + \sqrt{U^2 + 4n_i^2}}{2} \qquad (2.24)$$

Assuming that there is a differentiable relation between the electrically active concentration N and the total chemical concentration C, the expression for flux can be written as

$$J = -D_{eff}\nabla C \pm \frac{-DN}{\sqrt{U^2 + 4n_i^2}}\nabla U \qquad (2.25a)$$

$$D_{eff} \equiv D\frac{dN}{dC} \qquad (2.25b)$$

D_{eff} is often referred to as the effective diffusivity and will in general be less than the true diffusivity D since only a portion of the total impurity concentration is mobile at high concentration due to clustering. The numerical solution of the diffusion equation is formulated by approximating the continuous concentration profile with its values on a network of nodes within the device boundaries. The grid structure chosen for this numerical simulation is rectangular with nonuniform spacing in both spatial directions. The finite-difference approximation of Eqs. (2.23) at a node surrounded by its four nearest neighbors [2.19] is

$$\frac{\partial C_o}{\partial t} = \sum_{m=1}^{4}[B_m(C_m - C_o) \pm E_m(U_m - U_o)] \qquad (2.26)$$

where B_m and E_m are coefficients related to the grid spaces and the electron concentration [2.20].

For approximating the time derivative $\partial C_o/\partial t$, it has been found that a three-level time discretization scheme is advantageous for treating nonlinear parabolic equations. As an approximation,

$$\frac{\partial C_o}{\partial t} \simeq \frac{C_o^{n+1} - C_o^{n-1}}{k} \qquad (2.27)$$

where k is the value of the time step and the superscripts $n + 1$ and $n - 1$ denote values at the future and previous time levels, respectively.

NMOS Transistor Simulation

This section presents an example of a complete NMOS transistor fabrication process. The field region oxide growth and boron redistribution are simulated analytically, making the assumption that the boron concentration in the field region is low enough that its diffusivity will be constant. The program is then switched to numerical mode where the arsenic source/drain regions are implanted and driven in. In numerical mode the arsenic clustering and impurity interaction are taken into account. The structure file is saved and is used both for two-dimensional plots and to create the complete transistor structure.

a) Analytic Simulation of a Fully-recessed Oxide Isolation Region

The input command file shown in Fig. 2.13 simulates the initial portion of an NMOS process through the definition of the polysilicon gate. The substrate is p-type with <100> orientation and 20 ohm-cm resistivity. The structure is 5 μm deep and 3 μm wide. The height is specified to be 1 μm which allocates space for the thermal oxide that will be grown. The horizontal grid has 0.1 μm spaces at the device edges, decreasing toward the center of the device. Nodes inside the region are automatically generated with a nonuniform but smooth (actually quadratic) distribution.

The program is started out in analytic mode, since only in analytic mode can the silicon substrate be etched or thermal oxide be grown. After the silicon substrate is etched, an 80 nm pad oxide, a 70 nm nitride layer, and a 2 μm photoresist layer are deposited and selectively etched. The field region is implanted with boron and subsequently oxidized for 3 hours at 1000 °C in wet oxygen. The nitride is then stripped and an unmasked enhancement implant is performed. The polysilicon gate is deposited and etched with sloping sides. After switching to the numerical mode, the current result is saved into a file by SAVE card for further simulation. The current result is plotted two-dimensionally by using a PLOT.2D card. Equi-concentration lines are plotted within the range specified by FIRST and LAST. The next SUPRA input example file reads the saved result and plots the structure (Fig. 2.14).

b) Numerical Source/Drain Simulation

The input command file shown in Fig. 2.15 performs the numerical source/drain simulation. The initial structure is defined by loading the structure file EX2AST. This will be a high concentration source/drain simulation, so the program is placed in numeric mode. Arsenic is implanted, using the field oxide and polysilicon gate to define where the arsenic enters the silicon. The drive-in is then performed for 30 minutes at 1000 °C in an inert ambient and the resulting structure saved in the structure file EX2NST. The discretized boundaries of the thermal oxide and silicon substrate used in the numerical solution are plotted in Fig. 2.15. The contours of constant arsenic concentration are plotted with a dashed line type to distinguish them from the boron contour. Note that the electric field generated by the arsenic tends to pull the boron into the source/drain region where it segregates into the thermal oxide.

The whole transistor structure can be generated when the output file EX2NST is read twice. The second LOAD card, however, needs a parameter REFLECT to convert the structure of Fig. 2.15 to a mirror image. SUPRA is also capable of extending or shrinking the right or left side of the original structure during loading.

```
 1... COMMENT      Example 2 - NMOS transistor simulation
 2... COMMENT      Analytic recessed field region simulation

 3... STRUCTURE    P-TYPE ORIENTATION=100 DEPTH=5 WIDTH=3 HEIGHT=1
...  +             RESISTIVITY=20
 4... X.GRID       H1=.1  H2=.1  WIDTH=2  N.SPACES=34
 5... X.GRID       H1=.1  H2=.1  WIDTH=1  N.SPACES=10
 6... Y.GRID       H1=.1  H2=.02 DEPTH=1  N.SPACES=15
 7... Y.GRID       H1=.02 H2=.2  DEPTH=5  N.SPACES=40
 8... END          Structure definition

 9... COMMENT      Start out in analytic mode
10... ANALYTIC

11... COMMENT      Silicon etch to recess field regions
12... ETCH         SILICON  THICK=.4  START=.7  END=0  ANGLE=54.7

13... COMMENT      Pad oxide deposition
14... DEPOSIT      OXIDE   THICKNESS=.08

15... COMMENT      Nitride deposition and field region mask
16... DEPOSIT      NITRIDE  THICKNESS=.07
17... DEPOSIT      PHOTO.RESIST  THICKNESS=2
18... ETCH         PHOTO.RESIST  START=.7  END=0  ANGLE=90
19... ETCH         NITRIDE  START=.7  END=0  ANGLE=90

20... COMMENT      Boron field implant
21... IMPLANT      BORON   DOSE=5E12   ENERGY=100

22... COMMENT      Strip photoresist
23... ETCH         PHOTO.RESIST  START=0  CORNER=0  END=3

24... COMMENT      Field oxidation
25... OXIDIZE      TIME=180  TEMPERATURE=1000  WET

26... COMMENT      Strip nitride
27... ETCH         NITRIDE  START=0  CORNER=0  END=3

28... COMMENT      Unmasked enhancement implant
29... IMPLANT      BORON  DOSE=1E11  ENERGY=75

30... COMMENT      Poly silicon deposition and gate mask
31... DEPOSIT      POLYSILICON  THICKNESS=.5
32... ETCH         POLYSILICON  START=2.3  CORNER=2.1  END=0

33... COMMENT      Switch into numerical mode and save the structure
34... NUMERICAL
35... SAVE         STRUCTURE=EX2AST::-25
```

Fig. 2.13 SUPRA   input   simulating   recessed-oxide
         isolation.

```
1... TITLE      Example 2 - NMOS transistor simulation
2... COMMENT    Two-dimensional analytic simulation plot

3... COMMENT    Load analytic solution
4... STRUCTURE
5... LOAD       STRUCTURE=EX2AST::-25
6... END        Structure definition

7... PLOT.2D    X.MAX=5  Y.MAX=2
8... BOUNDARY   SMOOTH
9... CONTOUR    BORON FIRST=1E15  LAST=1E17  RATIO=10  OXIDE
10... END       2-D plot

*** END SUPRA ***
```

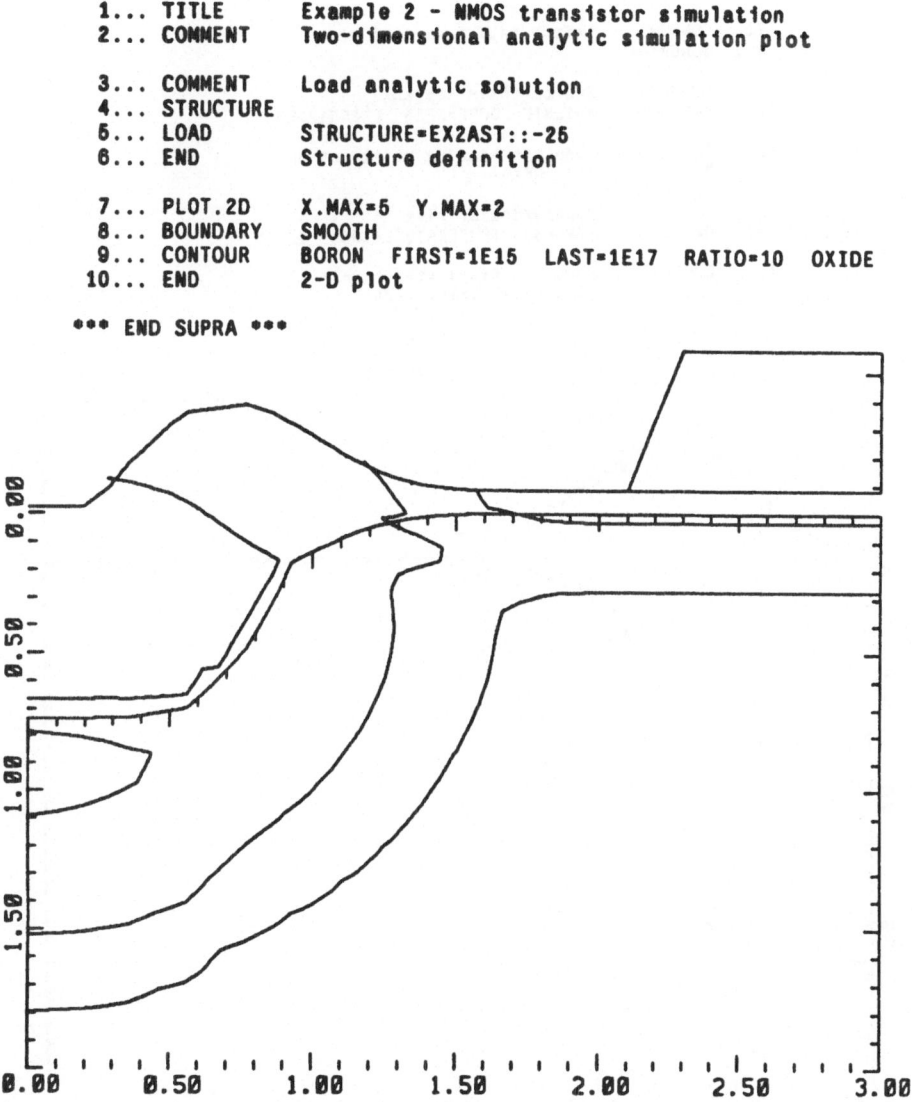

Fig. 2.14 SUPRA output showing a two-dimensional boron distribution near the channel-stop region.

```
1... TITLE        Example 2 - NMOS transistor simulation
2... COMMENT      Numerical source/drain simulation

3... COMMENT      Load analytic field region solution
4... STRUCTURE
5... LOAD         STRUCTURE=EX2AST::-25
6... END          Structure definition

7... COMMENT      Switch into numerical mode
8... NUMERICAL

9... COMMENT      Source/drain implant
10... IMPLANT     ARSENIC  DOSE=5E15  ENERGY=160

11... COMMENT     Turn the heading print-out on
12... PRINT       SET  HEAD

13... COMMENT     Source/drain drive-in
14... OXIDIZE     TIME=30  TEMPERATURE=1000   INERT

15... COMMENT     Save the final structure
16...' SAVE       STRUCTURE=EX2NST::-25
17... END
```

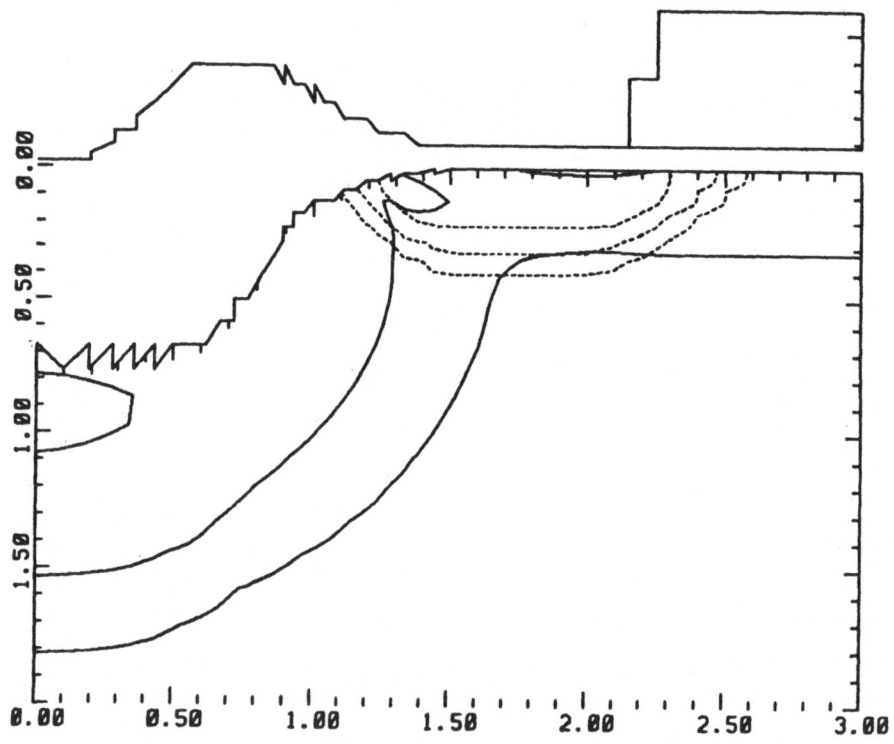

Fig. 2.15 SUPRA input and output simulating the
source/drain region.

2.4 SOAP : 2-D Oxidation Simulator

In today's VLSI technology, such extrinsic materials as polysilicon, nitride and oxide are deposited and patterned. The silicon substrate itself is etched to form a recessed surface. As a consequence, the silicon surface very often becomes uneven when a thermal oxidation process starts. The shapes of the oxide grown on such nonplanar surfaces are surprisingly different from those expected on a flat surface. For instance, the oxide on an etched silicon substrate forms a cusp that becomes thinner at an inside corner as shown in Fig. 2.16 [2.21]. At the outside silicon corner, the oxide thickness is also smaller than at the sidewalls. A similar phenomenon occurs when a gate oxide is grown in the vicinity of a thick field oxide edge [2.22]. This oxide thinning effect often causes a device failure because the breakdown voltage becomes lower near the corners. LOCOS is an excellent isolation technique that allows a self-aligned channel stop implantation and causes small parasitic

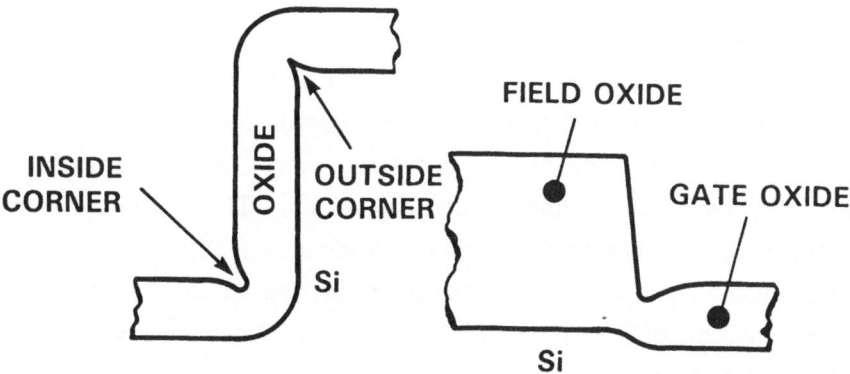

Fig. 2.16 Examples of non-planar oxidation.

capacitance. The isolation region takes almost half of the chip area and the transition region from the thin pad oxide under the nitride to the thick field oxide (so called bird's beak) becomes a non-negligible factor in a sub-micron transistor. Reduction of the bird's beak becomes one of the key issues in improving packing densities and device performance.

Two-Dimensional Oxidation Model

The two-dimensional oxidation model introduced in 1982 [2.23] is based on the physical behavior that the oxide can flow at high temperature. The viscous flow phenomenon was first observed in an experiment that a silicon wafer with thermal oxide on one side is heated in a nonoxidizing condition [2.24,2.25]. Since the thermal expansion of oxide is larger than silicon, the wafer bends toward the silicon substrate when the temperature is lower than 960 $^{\circ}$C. However, the wafer remains flat above the critical temperature, which suggests that stress due to the thermal mismatch is relaxed through viscous flow of oxide. The viscosity of oxide has been measured from the curvature of wafers as

$$\mu(T) = \mu_o \exp(E_\mu/kT) \qquad (2.28)$$

where μ_0 = 1.586E10 poises, and E_μ = 5.761 eV. This viscous flow is a dominant mechanism for the relaxation of stress induced by volume expansion during thermal oxidation. It should be noted that the viscosity increases tremendously as temperature goes down. Stress, therefore, may be insufficiently relaxed at temperatures below the glass-transition temperature (960 $^{\circ}$C) causing defect generation. The two-dimensional oxidation model in SOAP is based on this viscous flow [2.26].

As discussed in section 2.2, the oxidant diffuses in steady state. The flux conservation of Eqs. (2.3) and Eq. (2.4) is replaced by a generalized diffusion equation as

$$D_{eff}\nabla^2 C = \frac{\partial C}{\partial t} \simeq 0 \qquad (2.29)$$

where the effective diffusivity D_{eff} is assumed to be independent of oxidant concentration C or stress as in the linear-parabolic model. Boundary conditions for Eq. (2.29) are also obtained from the flux conditions as

$$\mathbf{F} \cdot \mathbf{n}(x,y) = kC \qquad\qquad \text{on} \quad S1$$
$$= - h[C_x - C(x,y)] \quad \text{on} \quad S2 \qquad (2.30)$$
$$= 0 \qquad\qquad\qquad \text{on} \quad S3, \ S4, \ S5$$

where $\mathbf{F}$ is the oxidant flux and $\mathbf{n}$ is a unit vector normal to the oxide surface (positive in the outward direction from the oxide bulk). The boundaries S1, S2 and S3 are the oxide/silicon interface, free-oxide surface, and nitride boundary, respectively, as defined in Fig. 2.17. These non-homogeneous conditions cause the oxidant distribution at the silicon/oxide interface to be non-uniform when a nitride layer is involved as in Eqs. (2.30) or the normal vector $\mathbf{n}$ changes direction. The oxidation rate, as a consequence, is different from place to place at the interface. The volume expansion rate, equivalent to the velocity of an oxide element close to the interface, can be given as

$$\mathbf{V}(x,y,t) = - (1 - \alpha)\frac{F(x,y,t)}{N_1}\mathbf{n}(x,y,t) \qquad \text{on} \quad S1 \quad (2.31)$$

where the minus sign indicates that the velocity is toward the oxide bulk and α is the ratio of the consumed silicon volume to the grown oxide volume. Every element in the already existing oxide bulk moves because of the newly grown oxide layer. The velocity, however, is very slow because oxidation rate is at most a few angstrom per second even in a steam ambient.

From the unique slow-viscous flow behavior of oxidation the motion of oxide elements can be described by a simplified Navier-Stoke's equation as

$$\mu\nabla V^2 = \nabla P \qquad\qquad (2.32a)$$

$$\nabla \cdot \mathbf{V} = 0 \qquad\qquad (2.32b)$$

where μ is the oxide viscosity, $\mathbf{V}$ is the velocity, and P is the pressure. The first equation (2.32a) indicates that the viscous damping force is

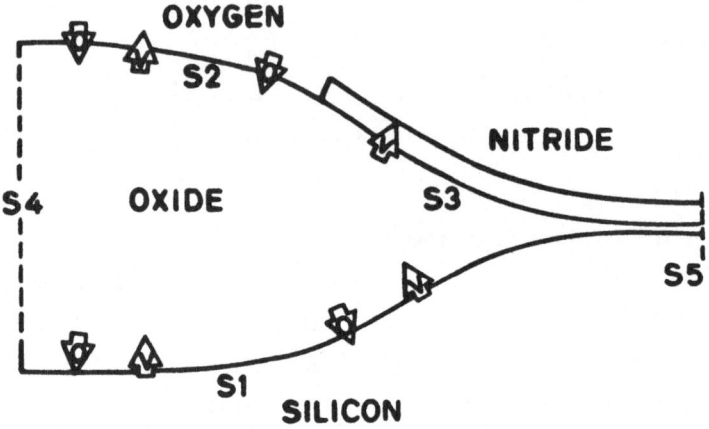

Fig. 2.17 Boundary conditions of the 2-D oxidation.

balanced with pressure gradient. Since the compressibility of oxide is very small (2.7E-12 cm^2/dyne), the oxide is assumed to be incompressible. This incompressibility characteristics requires that the density remain constant with respect to time. The continuity equation, consequently, is approximated below as

$$-\frac{d\rho}{dt} = \rho\nabla \cdot \mathbf{V} \simeq 0 \qquad (2.33)$$

where ρ is the oxide density. The velocity obtained from the O_2 concentration in Eq. (2.31) serves as a boundary condition on S1. On the oxide surfaces, boundary conditions are derived from an assumption that the internal pressure is balanced with external stress. Namely P becomes the ambient pressure on S2 and the nitride stress on S3, added by surface tension.

In the two-dimensional oxidation system, the ultimate solution we are looking for is the oxide boundary position that moves continuously. This moving-boundary problem is extremely difficult to solve with a conventional numerical method based on finite-difference or finite-element. The SOAP program adopted a different approach known as a boundary-value technique in which grid points are located on the

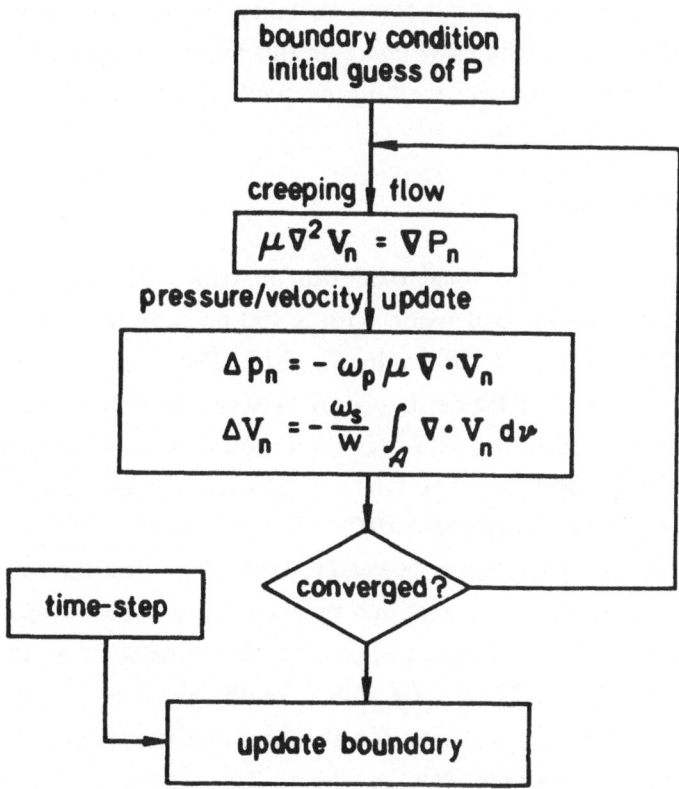

Fig. 2.18 Flow chart of the velocity-pressure iteration method.

boundaries and an integral form of Eqs. (2.29) and Eqs. (2.32) is solved by
using Green's function. Initially SOAP takes an initial guess that the
pressure distribution is constant and thus $\nabla P = 0$. The velocity field
calculated from this guessed pressure distribution would not satisfy the
zero divergence of Eq. (2.33) everywhere. The resulting $\nabla \cdot V$ is fed back
to obtain a new pressure distribution and then the velocity field is again
calculated from the new pressure. This velocity/pressure iteration
scheme continues until the incompressibility condition is satisfied within
an allowable range. The flow chart of this algorithm is given in Fig. 2.18.
Details of the numerical method as well as the physical model are
discussed elsewhere [2.26].

Application Examples

SOAP is applied to a local oxidation in which 0.6 µm field oxide is
grown from a 10 nm pad oxide with a 100 nm nitride layer. The input
file for this example is shown in Fig. 2.19. The starting oxide shape is
defined by several GRID cards placed between STRUCTURE and END
cards. A GRID card also allocates node points on a straight segment of
the initial oxide surface. The first GRID card assumes that the starting
position of the first segment is (0,0). The end position of the segment is
given by the two parameters (X.POSIT, Y.POSIT). The distances between
two adjacent nodes are determined by the total number of spaces within
the segment and the first/last spaces in the same way as in SUPRA.
However, to give sufficient flexibility to the segment orientation, two
different notations are allowed for the first and last spaces. Either HX1
or HY1, for an example, can be chosen for a segment that is neither
exactly in the horizontal nor in the vertical direction. If the segment is
only in the horizontal direction, HX1 and HX2 must be used because
there is no y-direction change. The boundary condition that the oxide
surface is facing is also specified in the GRID card. The present version
of SOAP accepts SILICON, NITRIDE, OXYGEN, OR REFLECT
boundary condition which specifies that the contacting material is a
silicon substrate, a nitride layer, oxygen ambient, or oxide bulk with
infinite extension, respectively. One should be cautioned that the position

```
Comment    Input file SOAP EX2 - Semi-Recessed Oxidation with Pressure
$                          Equilibration

Comment    Delimit lines of title with slashes; '*' => 'degree'
TITLE      /600 nm Semi-ROX/100 nm nitride/10 nm pad/950*C

Comment    Save data in file SOAP EX2DO (0 is default)
SAVE

Comment    Specify x and y plot bounds; display outline, title,
$          all contours, first and last nitride edge and nodes
PLOT.2D    X.MIN=0  X.MAX=2.0 Y.MIN=-0.5 Y.MAX=0.6 OUTLINE TITLE
+          CONT.ALL NIT.1ST NIT.LAST NOD.1ST NOD.LAST

Comment    Define structure (STRUCTURE and END required)
STRUCTURE
GRID       SILICON  X.POSI=2.000  Y.POSI=0.000  N.SP=20 HX1=.2   HX2=.2
GRID       REFLECT  X.POSI=2.000  Y.POSI=0.010  N.SP=1  HY1=.01  HY2=.01
GRID       OXYGEN   X.POSI=1.000  Y.POSI=0.010  N.SP=10 HX1=.2   HX2=.01
GRID       NITRIDE  X.POSI=0.000  Y.POSI=0.010  N.SP=10 HX1=.01  HX2=.2
+          THICK=0.15
END

Comment    Parameters for nitride plastic-elastic stress model; turn on
$          stress-dependent diffusion and reactivity
PHYSICAL   nit.len=.7  plastic=0.02 strdif.on strrea.on

Comment    Equilibrate pressure at each timestep (max 20 iterations);
$          set orientation effect on, set delta oxide = .05 nm
NUMERICAL  MAX=20 PRESS.CALC ORIENT DOX=.05

Comment    Oxidation parameters
OXIDIZE    TEMPERAT=1000  FOXIDE=0.6  IOXIDE=.010 WET BB.DELTA=.001




Comment    Input file SOAP EX2:1 - LOAD Output from Example 2
$          and Plot Stress Along Silicon Surface

Comment    Delimit lines of title with slashes; '*' => 'degree'
TITLE      /600 nm Semi-ROX/100 nm nitride/10 nm pad/950*C

Comment    Select type of 1D plot; lower (silicon) surface; specify plot
$          bounds; y axis spacing; specify outline,title,plot every other
PLOT.1D    STRESS LOWER  LEFT=0.0 RIGHT=2.0 BOTTOM=-2.0E09 TOP=3.2E09
+          YMAJ.SP= 1.0E09 YMIN.SP=0.1E09 OUTLINE TITLE DELTA.IN=2

Comment    LOAD file SOAP EX2DO (SAVEd in SOAP EX2); no continuation
LOAD       no.cont
```

Fig. 2.19 SOAP input file simulating a semi-recessed oxide.

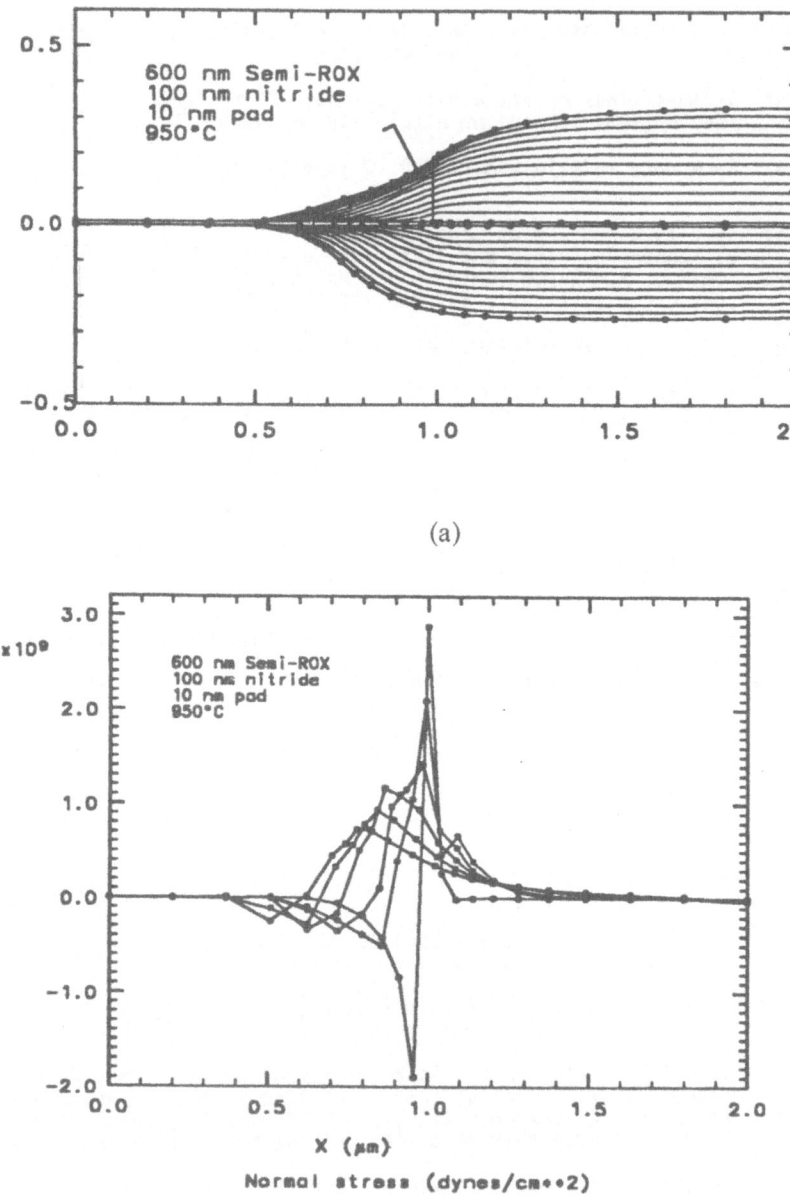

Fig. 2.20 SOAP output. (a) semi-recessed oxide shape
(b) stress distribution at the oxide/silicon
interface.

specified by the last GRID card is connected to the starting point (0,0) with a reflecting boundary condition.

The following NUMERICAL card is to control the numerical convergence, simulation time step, and orientation effect. The parameter DOX = 0.05 indicates that the incremental oxide thickness between two simulation steps is fixed by 0.05 μm and therefore the time step is determined by the oxide thickness. When NO.PRESS is specified in the NUMERICAL card, the program is forced to iterate only once in the pressure-velocity iteration algorithm. Although this mode gives an approximate solution that the pressure is constant, fairly close oxide shapes can be obtained with much shorter computation time. Therefore, it is recommended to use this NO.PRESS mode to check the current input file including the definition of the initial structure as well as the final oxide shape expected. The PHYSICAL card allows users to be able to change such parameters related to their own process conditions as oxide growth rate constants. If a user wants to see intermediate oxide shapes during calculation, he can specify a PLOT.2D card beforehand. The OXIDIZE card is to initiate an oxidation step. The final oxide thickness or the total oxidation time can be specified. The initial oxide thickness should be consistent with the initial oxide structure defined by GRID cards. The simulation result is shown in Fig. 2.20.

Reference

[2.1] D. A. Antoniadis, S. E. Hansen, R. W. Dutton, and A. G. Gonzales, "SUPREM I - A Program for IC Process Modeling and Simulation," SEL 77-006, Stanford Electronics Laboraties, Stanford University, Calif., May 1977.

[2.2] D. A. Antoniadis, S. E. Hansen, and R. W. Dutton, "SUPREM II - A Program for IC Process Modeling and Simulation," TR 5019.2, Stanford Electronics Laboratories, Stanford University, Calif., June 1978.

[2.3] C. P. Ho and S. E. Hansen, "SUPREM III - A Program for IC Process Modeling and Simulation," TR SEL 83-001, Stanford Electronics

Laboratories, Stanford University, Calif., July 1983.

[2.4] D. Chin, M. R. Kump, and R. W. Dutton, "SUPRA : Stanford University PRocess Analysis Program," Stanford University Laboratories, Stanford University, Stanford, Calif., July 1981.

[2.5] D. Chin, M. R. Kump, H. G. Lee, and R. W. Dutton, "Process Design Using Two-Dimensional Process and Device Simulators," *IEEE Trans. on Electron Devices* ED-29, Feb. 1982, pp. 336-340.

[2.6] D. Chin and R. W. Dutton, "SOAP : Stanford Oxidation Analysis Program," Stanford University Laboratories, TR SEL 83-002, Stanford University, Stanford, Calif. Aug. 1983.

[2.7] B. R. Penumalli, "A Comprehensive Two-Dimensional VLSI Process Simulation Program - BICEPS", *IEEE Trans. on Electron Devices* ED-36, Sept 1983, pp. 986-992.

[2.8] G. E. Smith, III, and A. J. Steckl, "RECIPE - A Two-Dimensional VLSI Modeling Program," *IEEE Trans. on Electron Devices* ED-29, Feb. 1982, pp. 216-221.

[2.9] K. A. Salsburg, and H. H. Hensen, "FEDSS - Finite-Element Diffusion - Simulation System," *IEEE Trans. on Electron Devices* ED-30, Sept 1983, pp. 1004-1011.

[2.10] J. Lindhard, M. Scharff, and M. Schiott, *Mat. Fys. Medd. Dan. Vid. Sclsk.* (33), 1963.

[2.11] B. E. Deal and A. S. Grove, "General Relationship for the thermal Oxidation of Silicon," *J. Appl. Phys,* 36(12), Dec 1965, pp. 3770-3778.

[2.12] C. P. Ho, J. D. Plummer, B. E. Deal, and J. D. Meindl, "Thermal Oxidation of Heavily Phosphorus Doped Silicon," *J. Electrochem. Soc.,* 125, Apr 1978, pp. 665-671.

[2.13] H. Z. Massoud, "Thermal Oxidation of Silicon in Dry Oxygen - Growth Kinetics and Charge Characterization in the Thin Regime," Stanford Electronics Laboratories, TR G502-1, Stanford University, Stanford, Calif., June 1983.

[2.14] R. O. Schwenker, E. S. Pan, and R. F. Lever, "Arsenic Clustering in Silicon," *J. Appl. Phys.,* 42, 1971, pp. 3195-3200.

[2.15] R. B. Fair and J. C. C. Tsai, "A Quantitative Model for the Diffusion of Phosphorus in Silicon and the Emitter Dip Effect," *J. Electrochem. Soc.*, 124, July 1977, pp. 1107-1121.

[2.16] H. Runge, "Distribution of Implanted Ions under Arbitrarily Shaped Mask Regions," *Phys. Stat. Sol. (a)*, vol. 39, 1977, pp. 595-599.

[2.17] J. Huang and L. Welliver, "on the Redistribution of Boron in the Diffused Layer during Thermal Oxidation," *J. Electrochem. Soc.*, vol. 117, 1970, pp. 1577-1580.

[2.18] H. G. Lee, R. W. Dutton, and D. A. Antoniadis, "On Redistribution of Boron during Thermal Oxidation of Silicon," *J. Electronchem. Soc.*, vol. 126, 1979, pp. 2001-2007.

[2.19] M. R. Kump and R. W. Dutton, "An Overview of Process Models and Two-Dimensional Analysis Tools," Stanford Electronics Laboratories, TR G-201-13, Stanford University, Stanford, Calif., July 1982.

[2.20] J. A. Greenfield and R. W. Dutton, "Nonplanar VLSI Device Analysis the Solution of Poisson's Equation," *IEEE Trans. on Electron Devices*, ED-27, Aug 1980, pp.1520-1532.

[2.21] R. B. Marcus and T. T. Sheng, "The Thermal Oxidation of Shaped Silicon Surfaces," *J. of Electrochem. Soc.*, 129, June 1982, pp. 1278-1289.

[2.22] L. O. Wilson, "Numerical Simulation of Gate Oxide Thinning in MOS Devices," *J. Electrochem. Soc.*, 129, Apr 1982, pp. 831-837.

[2.23] D. Chin, S. Y. Oh, S. M. Hu and J. L. Moll, "Two-Dimensional Modeling of Local Oxidation," presented at Device Research Conference, Colorado, June 1982.

[2.24] E. P. EerNisse, "Viscous Flow of SiO_2," *Appl. Phys. Lett.*, 30, 1977, pp. 290-293.

[2.25] E. P. EerNisse, "Stress in Thermal SiO_2 during Growth," *Appl. Phys. Lett.*, 35, 1979, pp. 8-10.

[2.26] D. Chin, S. Y. Oh, R. W. Dutton, and J. L. Moll, "Two-Dimensional Oxidation Modeling," *IEEE Trans. on Electron Devices*, ED-30, July 1983, pp. 744-749.

[2.27] D. Chin, S. Y. Oh, R. W. Dutton, and J. L. Moll, "Two-Dimensional Local Oxidation," *IEEE Trans. on Electron Devices*, ED-30, Sept 1983, pp. 993-999.

Chapter 3

Device Simulation

3.1 GEMINI : 2-D Poisson Solver

As the dimensions of MOS devices are scaled down, the device structures become more complicated. The insulator/semiconductor interfaces are often non-planar, and the impurity profiles of the devices are complicated and may not be expressed accurately in Gaussian form. The increased complexity of the device structure is necessary for optimization of the device performance, such as minimizing the drain-induced barrier-lowering effects, or enhancing the device reliability, e.g., reducing the electric field at the drain of the MOSFET. Therefore, in the development of VLSI MOS technology, it is essential to be able to simulate the electrical characteristics of devices which have complicated structures. The GEMINI program provides this capability.

The GEMINI program was developed at Stanford University by Greenfield and Dutton [3.1] in 1980. The program performs nonplanar VLSI device analysis by solving the 2-D Poisson equation. The program can accept data from the SUPREM and SUPRA programs, which provides high accuracy in the impurity profile definition, essential for submicron device simulations. The following sections will present in more detail the structure of the program as well as the input format, and then a simple example. In Part B of this book, the use of the GEMINI program is presented in many case studies.

65

The Capability of GEMINI

Within the 2-D simulation system, the GEMINI program is linked to the SUPREM and SUPRA programs, as well as the PLPKG program for very flexible graphical output. SUPREM provides one dimensional vertical profiles such as the channel and source/drain profiles. This is very useful since the vertical profiles are usually non-Gaussian, an obvious example being source/drain profile. The lateral profile at the source/drain of the MOSFET is specified in the GEMINI input file in this case. If the 2-D process simulator SUPRA is used, then the whole device structure is specified, with the exception of the placement of the electrodes. The SUPRA-GEMINI combination is useful for the simulation of device structures with novel 2-D geometries such as the LDD structure [3.2]. The link between GEMINI and the PLPKG program allows the graphical output of quantities such as potential, impurity, carrier concentrations and electric field distributions, with very flexible format. 1-D, 2-D, as well as bird's-eye-view from different angles are possible. Fig. 3.1 shows schematically the linking of the GEMINI program with the other programs within the 2-D simulation system.

The GEMINI program can simulate device structures such as MOSFET, JFET, MESFET, SOS devices, and other non-planar insulator/semiconductor structures such as the trench isolation structure (this will be described in Chapter 9). Fig. 3.2 shows three examples of structures whose electrical characteristics can be simulated by GEMINI. The structures shown are generated by the GEMINI program. The first one is a "channel length" simulation, where the device structure along the channel is defined. This is the structure that is most often simulated because it calculates the short channel effects where analytical or 1-D approximations are inadequate. The second example is a channel width simulation where narrow width effects such as threshold shifts are simulated. In this case, the capability of simulating non-planar insulator/semiconductor interface is essential. The third example shows an extreme case of a non-planar structure which is the trench isolation in

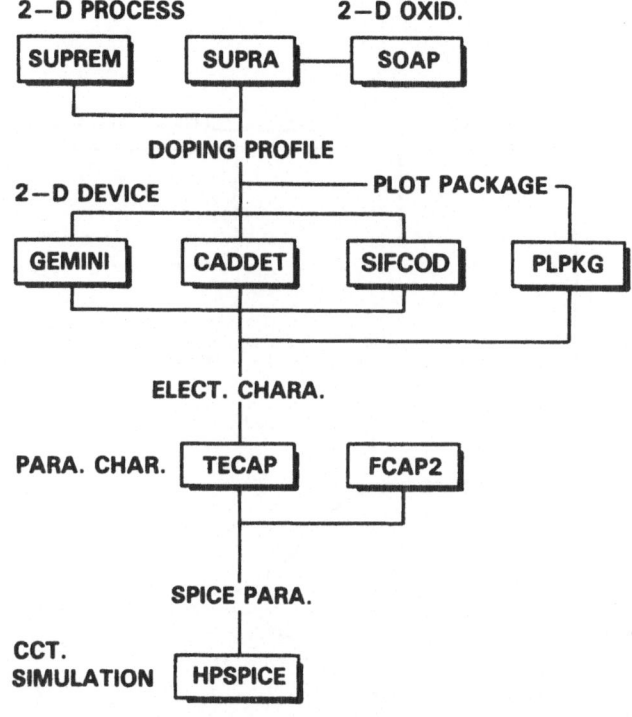

Fig. 3.1 Linkage of GEMINI in 2-D system

CMOS [3.3][3.4]. Here the trench is 5 μm deep and 1 μm wide. All of these simulations will be presented in detail in Part B of this book.

The program can extract device parameters such as the threshold voltage, subthreshold slope, punchthrough voltage, body effects, and electrode capacitance. Many other applications are also possible and the reader is referred to Part B of this book and also to reference [3.1]. GEMINI has graphic capabilities which allow 2-D plots of the device structure, junctions, depletion edge, as well as 1-D and 2-D contours of quantities such as potential, electric field, carrier concentrations and impurity concentrations. For details, the reader should refer to the GEMINI manual.

Basic Theory

The GEMINI program solves the 2-D Poisson equation only. The

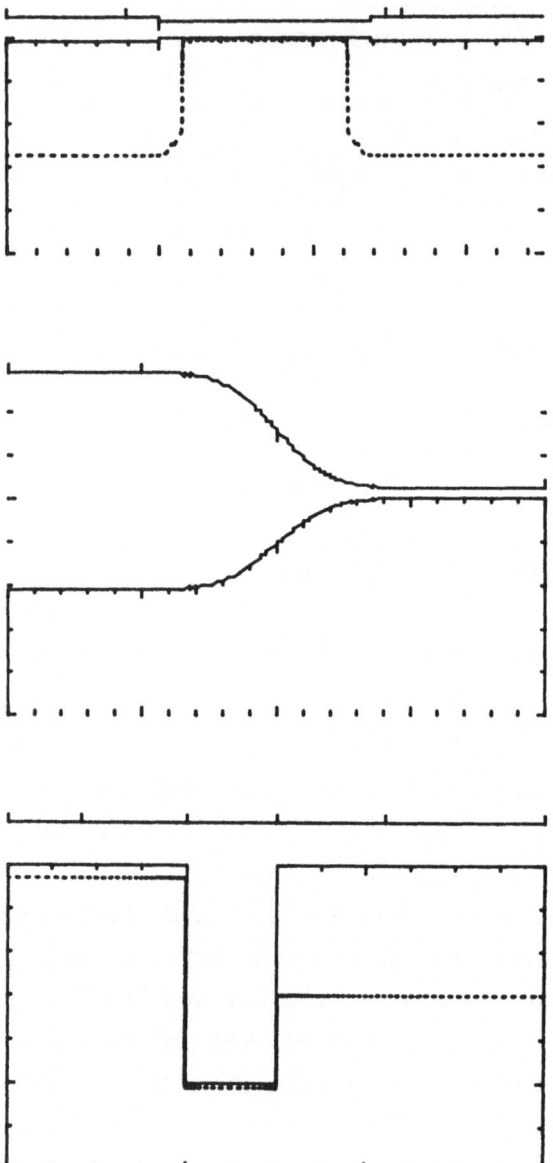

Fig. 3.2 Three major modes of GEMINI inputs.

Poisson   equation   is   expressed   by:

$$\nabla \cdot (\nabla\psi) = -\frac{q}{\varepsilon}(N+p-n) \qquad (3.1)$$

where ∇ is the gradient operator, ε is the dielectric constant, ψ is the potential, q is the electronic charge, N is the difference between donor and acceptor concentrations, and p and n are the hole and electron concentrations respectively. The electron and hole concentrations are related to the potential and quasi-Fermi potential by the Fermi-Dirac integral of order 1/2. In the program, the potential is defined such that it is equal to the substrate bias in the neutral substrate.

GEMINI does not solve the current continuity equations and it assumes constant quasi-Fermi level within each impurity region. For the MOSFET simulation, the majority carrier quasi-Fermi levels are set equal to the bias in the regions. For the minority carriers, the electron quasi-Fermi potential is set equal to the highest terminal bias and the hole quasi-Fermi potential is set equal to the lowest terminal bias in silicon. Hence the carrier concentration above the threshold voltage is inaccurate for large drain biases. The GEMINI program is useful for the linear   and subthreshold regions of the device characteristics.   The current calculations in these regions are outlined below.

In the linear region, where the drain to source bias (V_{DS}) is small (<0.1 V), the drain to source current (I_{DS}) is given by the product of the conductance of the channel (G_{DS}) and V_{DS}. G_{DS} is independent of V_{DS} since the current results mainly from minority carrier drift in the presence of the longitudinal electric field induced by the drain bias. The GEMINI program calculates the conductance by assuming zero drain to source bias. The current calculation in this case is valid for arbitrary gate bias. Knowing the device width W and mobility μ, the conductance is calculated by solving for the potential distribution and then determining the channel carrier concentration. For the channel length simulation, the program calculates the quantity Q_L which is defined by

$$Q_L = 1/\{ \int [\int Q \, dy]^{-1} \, dx\} \qquad (3.2)$$

where Q is the minority carrier concentration, and x and y are the horizontal and vertical coordinates respectively. The current can then be determined by the user, using the expression

$$I_{DS} = q \, \mu \, W \, Q_L \, V_{DS} \qquad (3.3)$$

where q is the electronic charge. Similarly, for the calculation of the device width, the quantity Q_W is calculated by the program, where Q_W is defined by

$$Q_W = \int [\int Q \, dy] dx \qquad (3.4)$$

and the channel current can be calculated using the expression

$$I_{DS} = 2 \, q \, \mu \, W \, Q_W \, V_{DS} / L \qquad (3.5)$$

In the subthreshold region, the current either results from the drain-induced barrier lowering effect, where the applied bias at the drain reduces the barrier height between the source and the drain, or from the lowering of the surface potential by an applied gate bias. In this region, the current can either flow along the interface between the gate dielectric and substrate, or through the substrate. In any case, there exists a point P in the device along the current path where the potential extremum is located. At P the electric field along the direction of current flow vanishes and the current is due to minority carrier diffusion. In this case, the current can be calculated from the potential distribution near this point.

In the calculation of the potential distribution, the minority carrier concentration in the substrate is assumed to be negligible compared with the ionized impurity concentration. Hence in the simulation of the punchthrough behavior in the subthreshold region, the results are accurate only for low current levels. However, reasonable accuracy is obtained even when this condition is violated, because the carriers are localized near the current path. The potential distribution, from which the current is calculated, is mainly determined by the device geometry, impurity profile and the applied biases.

GEMINI calculates the quantity Q_B from which the current can be determined. Q_B is given by:

$$Q_B = \frac{Z^*}{L^*} \frac{n_i^2}{N_B} \exp[\frac{s(V_s - V_p)}{kT/q}] \tag{3.6}$$

where $s = 1$ and -1 for p and n-channel respectively, V_p is the barrier potential, V_s is the source bias voltage, n_i the intrinsic carrier density, N_B the bulk impurity concentration, and Z^* and L^* are the effective width and length of the region of the potential distribution near the point P which controls the current. Fig. 3.3 shows schematically the significance of these two quantities in the case where the current path is through the bulk due to punchthrough. The user can then calculate the drain to source current by using the expression:

$$I_{DS} = -q\, DW Q_B[1 - \exp(\frac{sV_{DS}}{kT/q})] \tag{3.7}$$

where W is the width of the device, D the minority carrier diffusion constant $(= \mu kT/q)$.

Grid Definition and Numerical Techniques

The grid structure used by GEMINI is rectangular and has nonuniform spacing both in the horizontal and the vertical direction. This grid definition is chosen for accuracy, minimum storage and easy generation. Fig. 3.4 shows a typical grid definition for a channel length simulation. The grid spacings can be defined to be small in regions where the potential changes most rapidly, such as the regions near the source, the drain, and also near the surface. A maximum of 99 grids are allowed for the horizontal as well as the vertical coordinates.

Basically, the Poisson equation is solved numerically by using a five point approximation. Fig. 3.5 shows the discretization of the Poisson

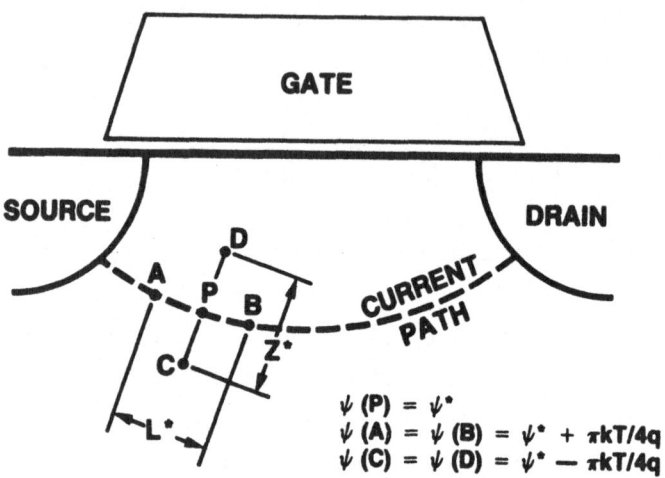

Fig. 3.3 Punchthrough current calculation.

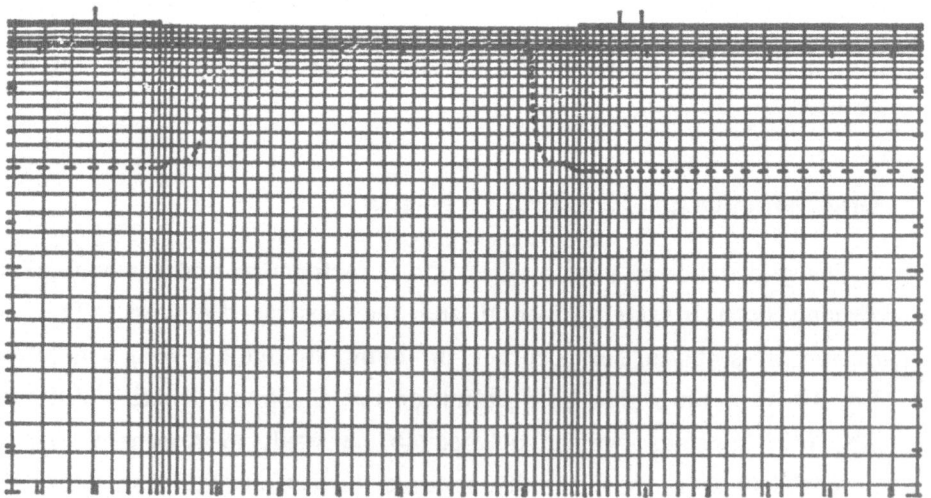

Fig. 3.4 Typical grid for channel-length simulation.

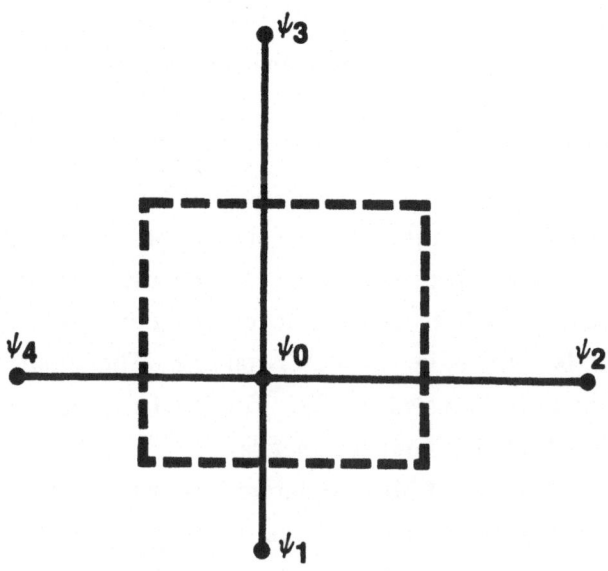

Fig. 3.5 Five-point finite-difference approximation

equation. A rectangular region is defined for each node point by the normal bisectors of the lines joining the node point to its four nearest neighbors. The component of the electric field normal to each side of this rectangle is taken to be constant along that side. The impurity concentration and free carrier concentrations are taken to be constant over the rectangle and are assigned the values obtained by evaluation at the coordinate of that node. For the interface and boundary conditions, the reader should refer to the GEMINI manual.

The discrete form of Poisson's equation consists of a nonlinear system of equations, one for each node point in the rectangular grid. This system of equations is solved by a one-step Newton-SLOR (Successive Line Over-Relaxation) iteration [3.5][3.6]. To determine if convergence is obtained, the program first locates the point along each vertical grid where the current path is most likely to be located. The potential at this point should be calculated most accurately. This point is defined as the point where the difference between the potential and the majority carrier quasi-Fermi potential is a maximum. The relative changes in the potential at these points are calculated. The iteration is terminated when all of the relative changes are smaller than the allowed value.

Example

In this example, a channel length simulation is presented and the subthreshold current will be calculated. In this case, the combination of SUPREM and GEMINI is used. The coupling between SUPRA and GEMINI will be shown in the next example. The input file for the GEMINI program is discussed briefly. The reader should refer to the manual for details of the input commands.

The input file for the GEMINI program is shown in Fig. 3.6. The input file begins with the structure definition. In this section of the input file, the temperature, device dimensions, substrate, grid, insulators, electrodes, interface charges, and the channel and source/drain profiles are all defined.

In the definition of the substrate, the impurity type and concentration has to be equal to that defined in the SUPREM input file for the simulation of the channel and source/drain profiles. The width and depth defines the region in which the Poisson's equation will be solved. The grid definition allows the user to specify the minimum and maximum grid spacings along the x-direction. The minimum spacing occurs at the left and right edges of the gate insulator region. The maximum spacing occurs at the center of the gate insulator region. The grid spacing changes from minimum to maximum parabolically. The grid spacing outside the gate region is then determined by the total number of horizontal grids specified and the number of grids used up in the gate region. The grid definition for the vertical coordinate consists of a minimum spacing definition, which is at the insulator/semiconductor interface and the total number of grids used for the vertical coordinate. These grid definitions allow the best accuracy in the determination of the potential distribution since more grids are available at the regions where the potential changes most rapidly.

```
COMMENT      NMOS CHANNEL LENGTH SIMULATION
STRUCTURE    TEMPERATURE=300
COMMENT      DEFINE SOLUTION REGION
SUBSTRATE    CONCENTRATION=6E14 P-TYPE WIDTH=5.38 DEPTH=4
COMMENT      DEFINE THE DISCRETIZATION GRID
GRID         XGRID.MIN=0.02 XGRID.MAX=0.05 X.SPACES=90
+            YGRID.MIN=0.01 Y.SPACES=60
COMMENT      DEFINE THE INSULATOR REGIONS
INSULATOR    SOURCE   THICKNESS=0.055 WIDTH=1 ENCROACH=1
INSULATOR    GATE     THICKNESS=0.0375
INSULATOR    DRAIN    THICKNESS=0.055 WIDTH=3 ENCROACH=1
COMMENT      DEFINE THE TOP SURFACE ELECTRODES
ELECTRODE    SOURCE ALUMINUM WIDTH=0.8
ELECTRODE    GATE    N+POLY WIDTH=1.5 LEFTEDGE=1.0
ELECTRODE    DRAIN   ALUMINUM WIDTH=2.8
COMMENT      DEFINE INTERFACE FIXED CHARGE
QSS          CONCENTRATION=2E10
COMMENT      DEFINE THE CHANNEL IMPLANT
PROFILE      CHANNEL IMPLANT SUPRMFILE=DL026
COMMENT      DEFINE THE S-D IMPLANT
PROFILE      SOURCE IMPLANT SUPRMFILE=DSDI5
+            WIDTH=1.0 X.CHAR=0.05
PROFILE      DRAIN  IMPLANT SUPRMFILE=DSDI5
+            WIDTH=3.0 X.CHAR=0.05
END
COMMENT      POISSON SOLUTION DEFINITION
SOLUTION     MAX.ITER=300 DATA.OUT=BD960B
BIAS         SUBSTRATE POTENTIAL=-1
BIAS         SOURCE POTENTIAL=0
BIAS         DRAIN POTENTIAL=3.0
BIAS         GATE POTENTIAL=0.0
END
STEP.VOL     GATE MAXIMUM DELTA.VOL=0.1 STEP=8
+            X.MIN=0 X.MAX=4 Y.MIN=0 Y.MAX=3
+            SUMM.OUT=BS960B
+            MAX.ITER=200
END
```

Fig. 3.6 Input file of GEMINI using SUPREM output.

The insulator defines the gate and source/drain region of the device. The thickness and width are defined. Note that GEMINI allows non-planar insulator-semiconductor interfaces. The ENCROACH parameter defines the width of the transition between the different insulator regions when the thicknesses are different. This is essential for the channel width simulation where the field isolation oxide structure has to be defined.

The electrode command defines the materials and positions of the electrodes where biases will be applied to the device. The source and drain electrodes automatically make contact with the diffusion regions. The electrodes cannot touch in the definition. Fixed charge density at the insulator-semiconductor interface can be specified by the QSS command.

The channel profile can be specified either in the GEMINI input file using Gaussian expressions, or by accepting data from a SUPREM simulation. The latter method is generally used, since the SUPREM simulation provides the best profile definition. The channel profile will be placed along the gate region, as well as the source and drain region, which is typically what is done in device fabrication. The source and drain profiles are then specified by first defining the width of these regions. Then the data file name of the SUPREM simulations are specified. Since SUPREM is a one dimensional profile, the lateral diffusion of the source and drain profiles have to be defined in GEMINI. In this case, the characteristic length X.CHAR of a lateral Gaussian distribution is defined. The structure definition for the device is ended by an END command.

The next section of the input file is the solution definition, where the bias conditions are specified. In this example, the substrate, source, drain, and gate biases are specified. If the device definition has an error which causes non-convergence, the MAX.ITER specifies the maximum number of iterations allowed. For a MOSFET simulation, the number of maximum iterations needed is typically 100 or less. The output data file name which contains the solution of the Poisson's equation are also defined. The END command ends the solution definition.

The next section of the input file is for parameter extraction. The subthreshold characteristics of the device can be generated by stepping the gate bias and calculating the quantity Q_B, from which the current can be determined. The STEP.VOL definition specifies the voltage to be stepped, which is GATE in this example. The voltage increment DELTA.VOL and the number of steps STEP are specified. The MAXIMUM parameter defines the path where the current will flow.

For this example, the carriers are electrons. Hence the current path is located where the potential is a MAXIMUM. For a p-channel MOSFET simulation, the parameter would be MINIMUM. The approximate region where the current path will be located is specified to minimize the search

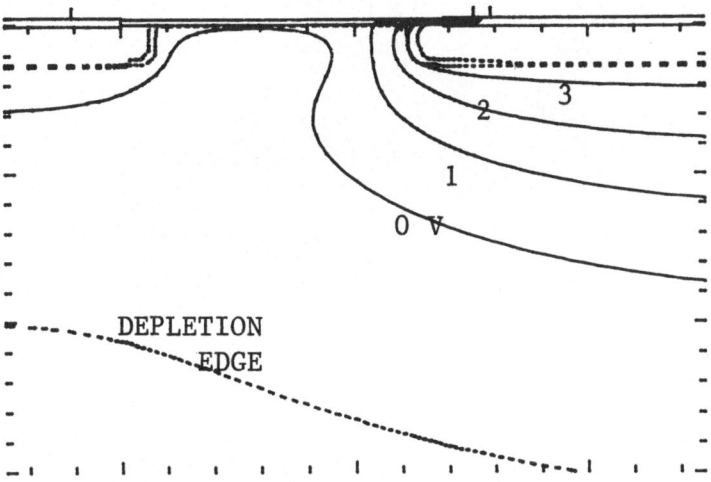

Fig. 3.7 Device structure and 2-D potential distribution
of the example.

time for the current path. The summary output file name is specified here. The summary consists of the applied voltages as well as the quantity Q_B as a function of the gate voltage. The END command ends the STEP.VOL definition.

With the SUPREM data files available and the GEMINI input file completed, the program can be executed and the solution plotted in various formats.

Fig. 3.7 shows the 2-D plot of the device structure and also the potential profiles with the device biased as defined in the SOLUTION definition. By using the PLPKG program, the bird's eye view of the impurity and potential profiles are plotted and shown in Fig. 3.8 and 3.9. The calculated device subthreshold characteristics are plotted in Fig. 3.10. From this graph, the subthreshold slope and threshold voltage (as defined by a threshold current) can be determined.

The second example briefly shows the results of the SUPRA and

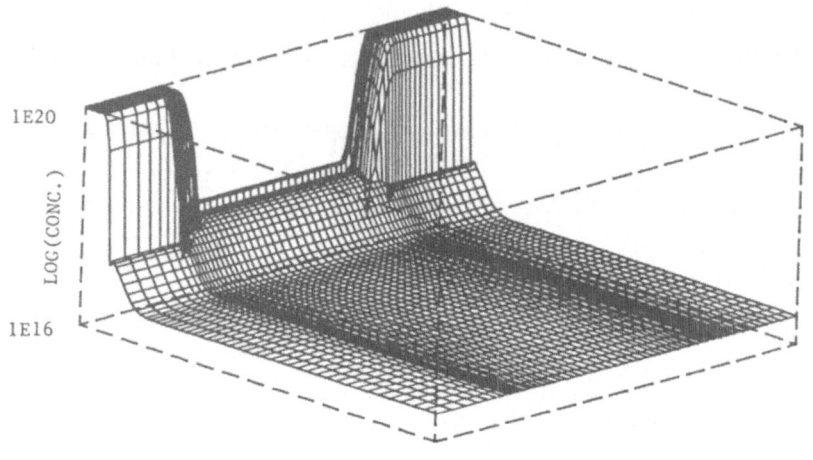

Fig. 3.8 Bird's-eye-view of impurity distribution.

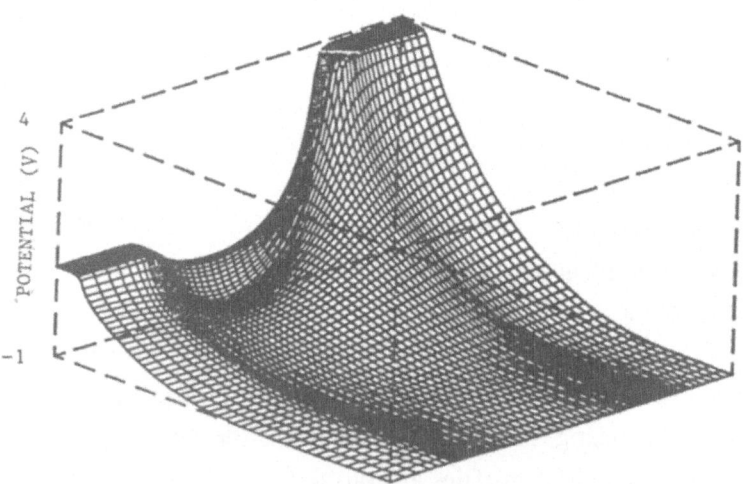

Fig. 3.9 Bird's-eye-view of potential distribution.

GEMINI combination in the channel width simulation. The reader is referred to Chapter 10 for more details of using SUPRA-GEMINI for calculating narrow width effects. Fig. 3.11 shows the input file for the GEMINI program. In this case, the input file is very simple since the structure is essentially defined by the SUPRA data file. The only structure input in this example is the gate definition. The SOLUTION and STEP.VOL definitions are the same as the SUPREM-GEMINI combinations. Fig. 3.12 shows the structure of the device, including the 2-D impurity profile, and also the depletion edge calculated by the GEMINI program. One disadvantage in the SUPRA-GEMINI combination is that the grid is defined as in the SUPRA simulation, which is often not optimized for potential calculations. Fig. 3.13 shows the I_{DS}-V_{GS} characteristics of this device simulated by the GEMINI program. This type of channel width simulation is very useful for the development of isolation technologies where the narrow width effects are minimized.

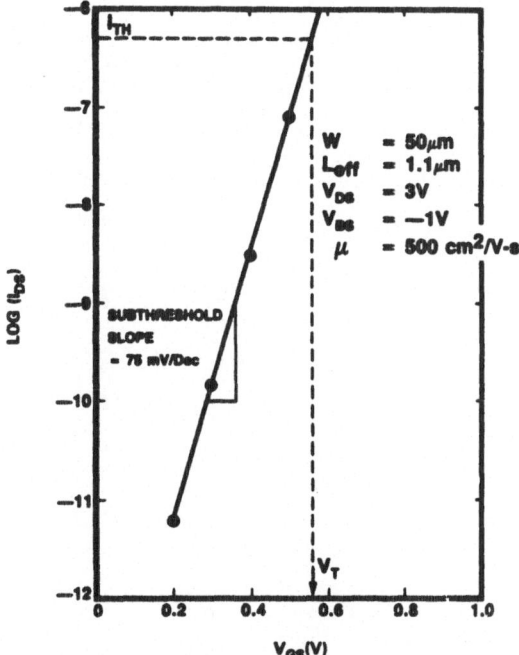

Fig. 3.10 N-channel MOSFET Subthreshold
characteristics simulated by GEMINI.

```
COMMENT LOCOS ISOLATION CHANNEL WIDTH SIMULATION
STRUCTURE DATA.INP=GSW76
ELECTRODE GATE N+POLY  WIDTH=0.99  LEFTEDGE=1
END
SOLUTION  MAX.ITER=300  DATA.OUT=LOS78
BIAS      SUBSTRATE  POTENTIAL=0
BIAS      GATE       POTENTIAL=1
END
```

Fig. 3.11 GEMINI Input file for the channel width
     simulation.

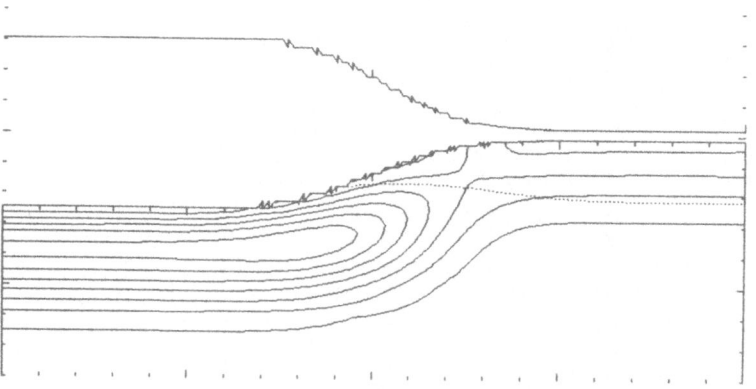

Fig. 3.12 Device structure of channel width simulation.

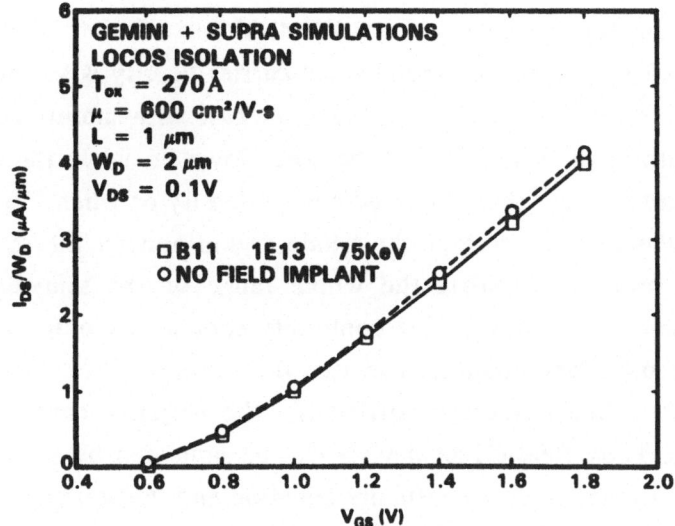

Fig. 3.13 Linear region characteristics for the narrow
width devices.

3.2 CADDET : 2-D 1-Carrier Device Simulator

As shown in the previous section, GEMINI can simulate a MOSFET when either V_{DS} is small or the channel carrier density is so small that potential distribution is not disturbed as is the case with subthreshold or a weak punchthrough region. It can't be used, however, when the channel carrier density is larger than the fixed charge density or when V_{DS} is not small as is the case in most of the triode region and all the saturation region. In order to simulate the whole range of the semiconductor device, Poisson, electron and hole continuity equations should be solved with proper boundary conditions in two dimensions. In the field-effect device, most of the current is carried by the majority carrier of the source/drain. Thus, field-effect devices can be simulated by solving only the majority carrier current continuity equation and Poisson's equation to save calculation time. CADDET is a dedicated program for the field-effect transistor and functions in this way, so it can simulate the whole range of the field-effect transistor (subthreshold, linear, saturation and punchthrough) with reasonable CPU time.

CADDET can analyze the planar JFET's and MOSFET's with various channel, source, and drain configurations. Fig. 3.14 illustrates the input structures of the field-effect transistors that can be simulated by CADDET. The only limitation is that the silicon and oxide surface should be planar. The device structure is selected by the structure name and the impurity profiles are specified by the step and Gaussian profiles. For these field-effect device structures, the whole operation can be simulated and the terminal currents can be calculated without any limitation on bias. CADDET has been linked to the PLPKG program so that the distributions of the electron, hole, impurity, and potential can be displayed in 1-D, 2-D and 3-D plots. The distribution of the field and velocity can also be plotted. CADDET is widely used to analyze the velocity saturation, break-down, hot electron generation, punchthrough, and the LDD structure. An analysis of the LDD structure will be presented in detail in Part B. CADDET is also indispensable for generating the whole I-V characteristics in order to extract SPICE

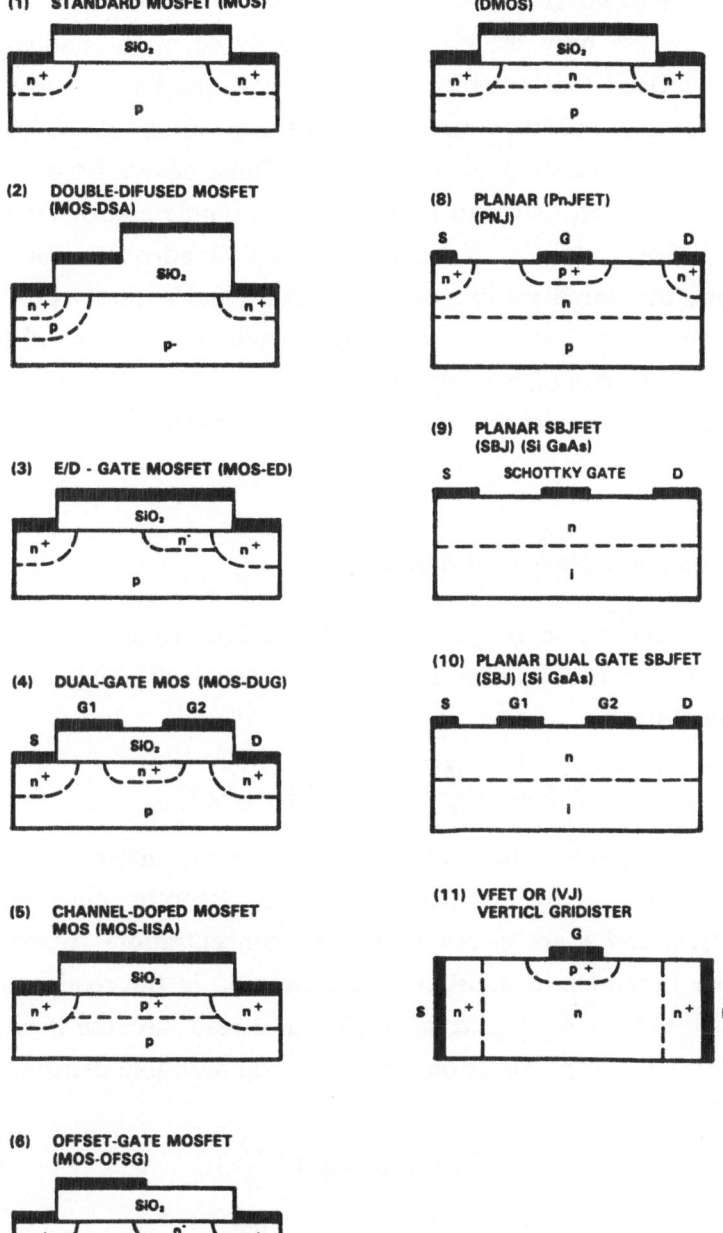

Fig. 3.14 Input structures of CADDET.

parameters for the circuit simulations. There are several limitations in
CADDET due to the simplifications. First, only n-channel devices can be
simulated because only the electron continuity equation is solved. For
p-channel devices, the structure should be converted to the equivalent
n-channel device and then simulated. The simulation results are then
converted to the p-channel characteristics. Those device structures with
a planar surface and Gaussian profile are the simple approximations of
the real device structures. Though CADDET is adequate for many
applications, the simplifications of the structure and impurity profile can
be a source of inaccuracy in scaled submicron devices, where the
effects of the non-planar structure and non-Gaussian profile are
significant. For more details of CADDET, please refer to the CADDET
manual [3.7].

Basic Equation and Numerical Algorithm

To cover the whole range of the field-effect transistor operation,
CADDET solves the Poisson and electron continuity equations. The
Poisson equation is

$$\nabla^2 \psi = -\frac{q}{\varepsilon}(p - n + N_D^+ - N_A^-) \tag{3.8}$$

where ψ is the electrostatic potential and ε is the dielectric constant;
N_D^+ and N_A^- are ionized donor and acceptor concentrations
respectively; n and p are electron and hole concentrations respectively.
Though the Fermi-Dirac statistics should be used in the semiconductor,
the Maxwell-Boltzman statistics is employed instead, because it is simple
and is still a good approximation. Then electron and hole density can be
expressed by

$$n = n_i \exp[(\psi - \phi_n)/\frac{kT}{q}] \tag{3.9a}$$

$$p = n_i \exp[(\phi_p - \psi)/\frac{kT}{q}] \tag{3.9b}$$

where n_i is the intrinsic carrier density; T is the absolute temperature,
and k is the Boltzman constant; ϕ_n and ϕ_p are the quasi-Fermi potentials

of electrons and holes respectively.

The electron continuity equation is formulated using the stream functions [3.8] as a basic variable instead of electron and hole concentrations because of the numerical stability. When the recombination and generation of carriers are neglected, which is reasonable in MOSFETs, the divergence of the electron current density is zero. It can be a curl of some vector (stream function). The current density (J) is formulated as below when the normalization ($kT/q = 1$) and Einstein's relation are used.

$$\nabla \cdot \mathbf{J}_n = 0 \qquad (3.10a)$$

$$\mathbf{J}_n = \mu_n \, e^{\psi} \, \nabla(ne^{-\psi}) \qquad (3.10b)$$

$$\mathbf{J}_n = J_o \nabla \times \theta \qquad (3.10c)$$

For a two-dimensional case, the stream function has only the z component, θ. The x and y components of electron current density can be written

$$J_x = \mu_n \, e^{\psi} \frac{\partial}{\partial x}(n \, e^{-\psi}) = J_o \frac{\partial \theta}{\partial y} \qquad (3.11a)$$

$$J_y = \mu_n \, e^{\psi} \frac{\partial}{\partial y}(n \, e^{-\psi}) = -J_o \frac{\partial \theta}{\partial x} \qquad (3.11b)$$

Divide both sides of the above equations by $\mu_n e^{\psi}$, then differentiate the first equation partially with respect to y, the second equation partially with respect to x, and add the two obtaining

$$\frac{\partial}{\partial x}[\mu_n^{-1} \, e^{-\psi} \frac{\partial \theta}{\partial x}] + \frac{\partial}{\partial y}[\mu_n^{-1} \, e^{-\psi} \frac{\partial \theta}{\partial y}] = 0 \qquad (3.12)$$

This equation is equivalent to the current continuity Eq. (3.10a). When θ is solved with appropriate boundary conditions, J_o should be evaluated using the following equation based on Eq. (3.11a).

$$J_o = \frac{- N_D[1 - e^{- V_{DS}}]}{\displaystyle\int_0^{L_{eff}} \mu_n^{-1} e^{-\psi} \frac{\partial \theta}{\partial y} dx} \qquad (3.13)$$

The integration in the denominator is along the channel. The electron density is calculated using Eqs. (3.11a) and (3.11b) with J_O and Θ.

Both equations are discretized by the standard 5-point finite-difference approximation. A non-uniform grid is employed to enhance accuracy. The mesh is automatically generated by the program as shown in Fig. 3.15. In the vertical (y) direction, a fine grid is used at the silicon surface and grid spacing increases geometrically toward the substrate. Fine grids are used in the channel and source/drain junctions and coarse grids are used in the middle of the channel in the horizontal (x) direction. A user can specify the total number of grids, the minimum spacing and geometrical ratio for both x and y directions. The maximum number of grid points is limited to 2000. The meshes for potential and for stream function are different and are interleaved as shown in Fig. 3.16. The circle is the node for the potential and the cross is the node for the stream function. The stream function mesh is located in the middle of the potential mesh. To reduce the calculation time, these equations are solved iteratively (Gummel iteration) after discretization instead of by the simultaneous solution of both equations. The flow chart of the program is illustrated in Fig. 3.17. When the program starts, the grid is generated, the potential, hole and electron quasi-Fermi potentials are initialized, and the equations are discretized. The hole quasi-Fermi potential is set constant throughout the device and is the same as that of the substrate. First, the Poisson equation is discretized and solved by Stone's method [3.9] for the potential with the fixed electron quasi-Fermi potential. Then, the electron continuity equation is discretized and solved by SLOR(Successive Line Over Relaxation) for the electron stream function with fixed potential. The electron densities and electron quasi-Fermi potential are calculated from the stream function and are updated. This procedure is repeated until the solutions converge. Finally, the drain current is calculated from the converged stream function.

Mobility Model

In MOS devices, the drain current is directly proportional to the mobility. Because an accurate mobility model is very important to the

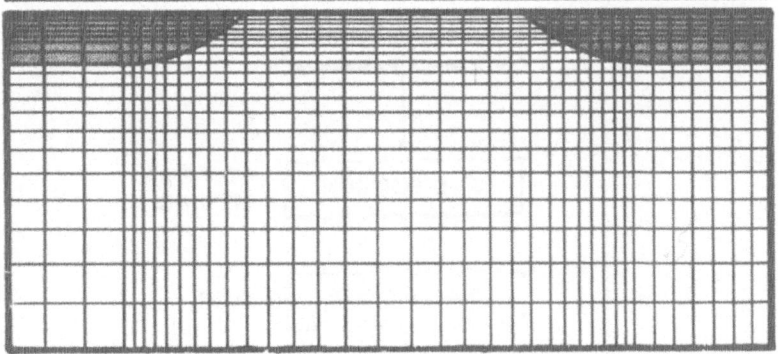

Fig. 3.15 Non-uniform rectangular grid in CADDET.

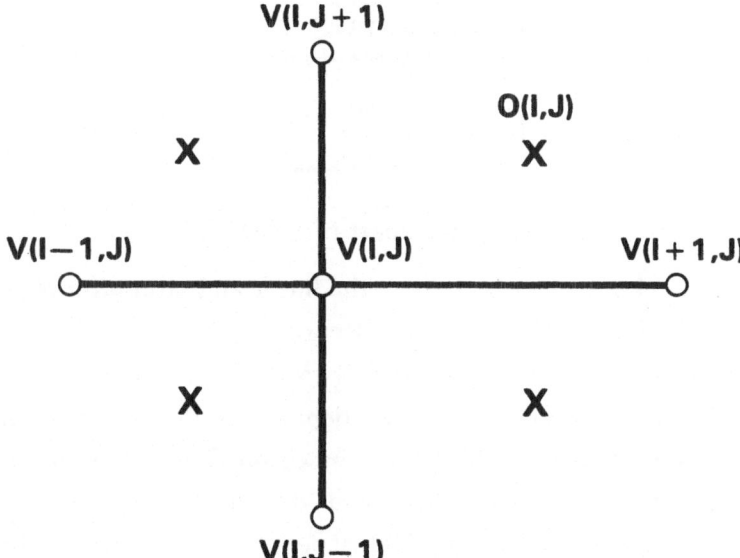

Fig. 3.16 Meshes of potential and stream functions.

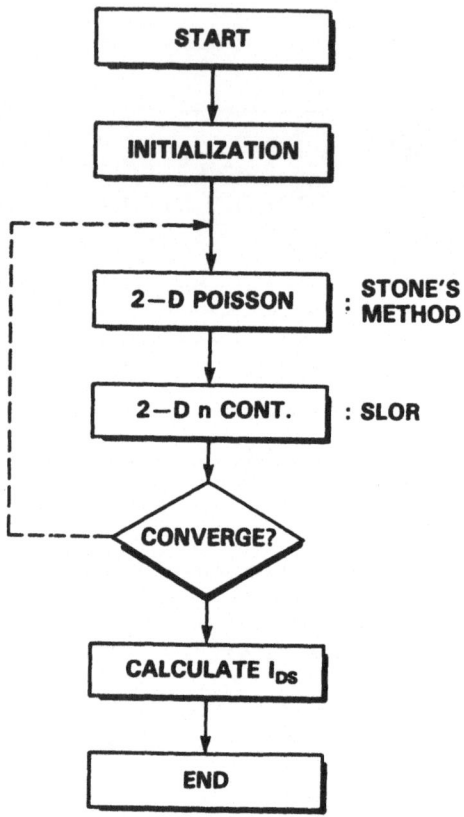

Fig. 3.17 Flow chart of CADDET.

drain current calculation especially in the linear and saturation regions, it is presented in detail in this section.

Mobility depends on both the parallel field and the doping density in the bulk silicon. In MOS devices, it also depends on the normal field. The original mobility model in CADDET is based on Gummel's bulk mobility model [3.10] for the parallel field and doping dependence, and the empirical model for the normal field dependence. The model equations are shown below [3.11]

$$\mu_n = \mu_o f(N_B, E_p) \, g(E_n) \qquad (3.14a)$$

$$f(N_B, E_p) = [1 + \frac{N_B}{N_B/S + N} + \frac{(E_p/B)^2}{N_B/S + N} + (\frac{E_p}{B})^2]^{1/2} \quad (3.14b)$$

$$g(E_m) = (1 + \alpha E_n)^{1/2} \quad (3.14c)$$

where N_B is the impurity density; E_p and E_n are the field components parallel and normal to the current-flow, respectively; μ_0 is 1400 cm^2/V sec, and S is 350; N is 3×10^{16} cm^{-3} and A is 3.5×10^3 V/cm; F is 8.8, and B is 7.4×10^3 V/cm; α is 1.54×10^{-5} cm/V. This model, however, doesn't agree well with the measurements, especially in the thin gate and the highly doped channel. In Fig. 3.18, simulations with this model are compared with the measurements. The thickness of the gate oxide is 25 nm and the effective channel length is 0.9 μm. The drain voltage is 0.1 volts in Fig. 3.18(a) and 5 volts in Fig. 3.18(b). Disagreement is severe in the saturation region, due to the way the normal field degradation is combined with the bulk mobility model. The normal field not only degrades the zero parallel field mobility but also the saturation velocity.

A new mobility model has been developed. It is based on a modified Gummel's model [3.12] and on the newly characterized zero-parallel field mobility [3.13]. Two equations are combined in such a way to keep the saturation velocity constant when the zero parallel field mobility is degraded to obey the Thornber's scaling rule [3.14]. The new model is

$$\mu_o(E_n, E_p) = \frac{\mu_o(E_n)}{[1 + \frac{\{\mu_n(E_n)E_p/V_c\}^2}{\mu_o(E_n)E_p/V_c + G} + \{\frac{\mu_o(E_n)E_p}{V_s}\}^2]^{1/2}} \quad (3.15a)$$

$$\mu_o(E_n) = 690 (E_n/10^5)^{-0.28} \quad (3.15b)$$

where V_s is 1.036×10^7 cm/sec and V_c is 4.9×10^6 cm/sec; G is 8.8. The normal field, E_n, is in volt/cm. The comparisons of the new model and the measurement are illustrated in Fig. 3.19. Agreements are good in the linear and the saturation regions. CADDET simulations with the new model have been bench-marked for a wide range of device parameters. Agreements are within 10%. CADDET with the new mobility model

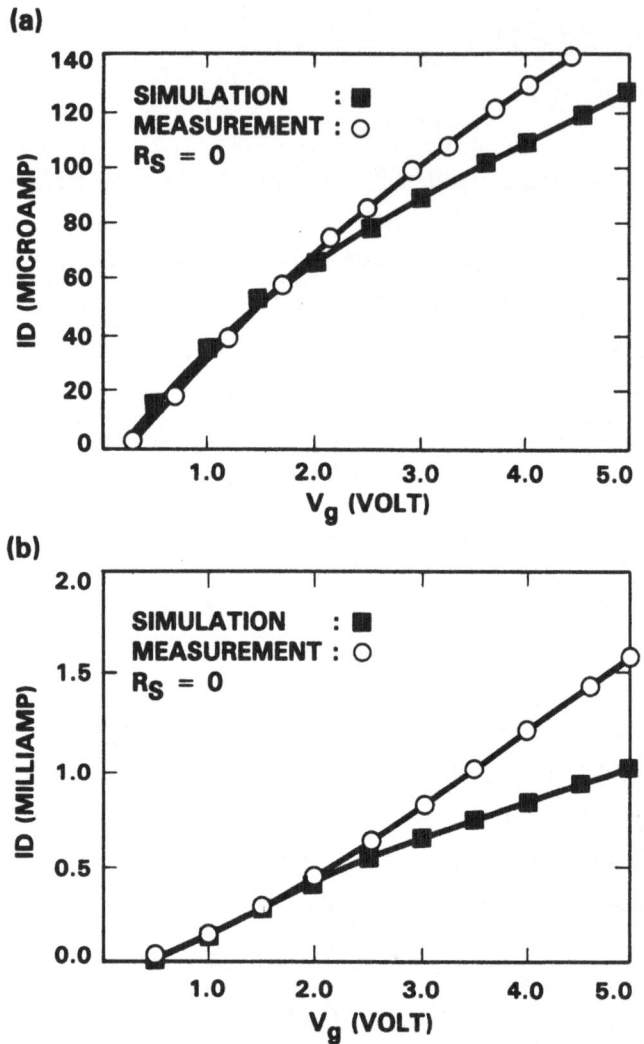

Fig. 3.18 Bench-mark   of   CADDET   mobility   model.
(a)$V_{DS} = 0.1$ V $(b)V_{DS} = 5$ V.

enables the accurate simulations of the whole range of MOS devices with
reasonably fast computation time. In the next section, the simulation of
the conventional MOS devices will be illustrated as an example of
CADDET applications.

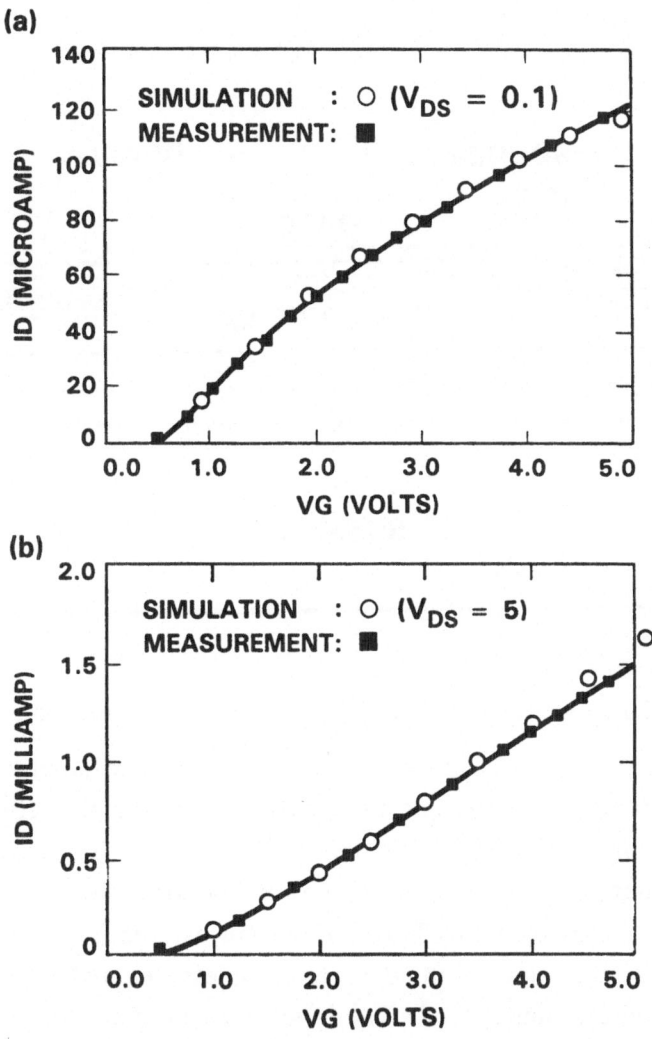

Fig. 3.19 Bench-mark of the new mobility model.

N-channel MOSFET simulation by CADDET

In CADDET, several MOSFET and JFET structures can be simulated such as conventional MOS, DMOS, p-n junction FET, Schottkey-barrier JFET and vertical FET as shown in Fig. 3.14. In this section, only the simulation of the conventional MOSFET will be illustrated as a simple case study to show how to use CADDET. Fig. 3.20 shows the structure of an n-channel MOSFET to be simulated. The gate oxide thickness is

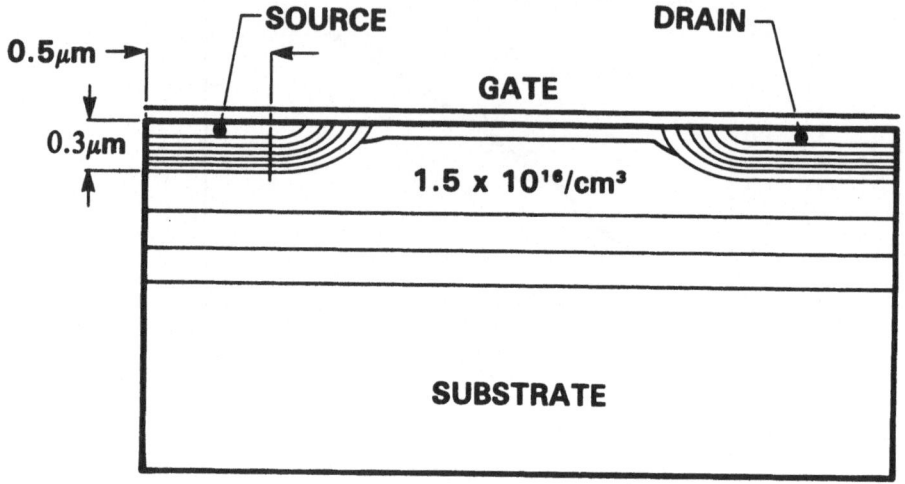

Fig. 3.20 Device structure of n-channel MOSFET example.

46.8 nm and the effective channel length is 1.0 μm. The junction depth of the source/drain is 0.3 μm and the substrate doping is 6×10^{14} cm^{-3}.

Before running the program, an input file must be prepared which describes the device structure and doping distribution. The input for the above device is shown in Fig. 3.21. A fixed format input is employed. The problem of this kind of input is that it is easy to make errors and difficult to read and understand. The input consists of 80 column cards. A card is divided into fields. The first card   is divided into 5-column fields and the rest is divided into 10-column fields. One feature of CADDET that differs from other device simulators is in the unit system. CADDET uses the MKS system. That is, length is measured in meters, weight in kilograms, and time in seconds. Device type and other control parameters are specified in the first card. "MOS" in the first field designates the device type as a conventional MOSFET. The second field specifies the number of gates as one. The second card specifies the numbers of x and y grids and their spacings. The third card is the device parameters. Temperature is in the first field as 300 °K, mobility in the

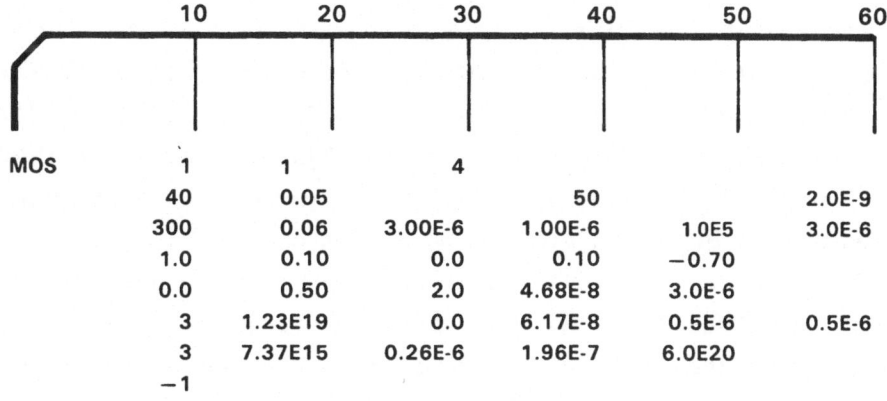

Fig. 3.21 CADDET input for the example device.

second field as 0.06 m^2/volt sec, device length in the third field as 3 μm, channel width in the fourth field as 1 μm, and the substrate depth in the sixth field as 3 μm. In the fourth card, the drain biasing schedule is described. The first parameter gives the back gate bias as -1 volt. In CADDET, the back gate bias is specified as V_{SB}, that is, the source bias with respect to the substrate. Be careful ! It is the negative of the conventional back gate bias. The second parameter is the starting drain bias. The stepping voltage and the final bias of V_{DS} are specified in the third and fourth fields, respectively. The drain bias is fixed at 0.1 volt in this example. The flatband voltage is in the fifth field. The fifth card is the gate biasing. The initial gate bias is in the first field, the voltage step in the second field, and the final gate bias in the third field. In this example, V_{GS} starts from 0 volts and stepped by 0.5 volts to 2.0 volts. The fourth and fifth parameters specify the gate oxide thickness and gate length, respectively.

The next card describes the doping distribution of the source/drain. In CADDET, the source/drain profile can be step, Gaussian or implanted Gaussian. The first parameter in the fifth card gives the doping type as implanted Gaussian. The shape and equation in the implanted Gaussian is illustrated in Fig. 3.22. The dose, range, and standard deviation are specified in the second, third and fourth fields respectively. The fifth and sixth parameters specify L_S and L_D respectively. The channel profile is in the seventh card. The first parameter is the types of the channel profile. It can be a uniform, a redistributed or an implanted profile. The implanted profile is selected in this case and illustrated in Fig. 3.23. The dose (N_p'), range (R_p') and standard deviation(σ') are in the second, third and fourth fields, respectively. The fifth parameter is the substrate doping (N_A). In the last card, "-1" signals the end of the input. After preparing the input, CADDET can be run either in the interactive or batch mode. When CADDET is run, it first asks for the input file name. Then it reads the input file and echoes the input to the terminal. After that, it starts to iterate to solve the Poisson and electron continuity equation. It prints out the results of each iteration. The drain currents and bias conditions are summarized and saved in the summary file as shown in Fig. 3.24.

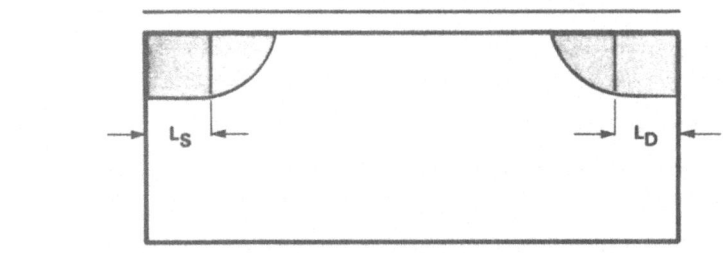

$$
N_i\,(i,j) = \begin{cases} \dfrac{N_p}{\sqrt{2\pi}\sigma}\,\exp\left[-\dfrac{(y_j - R_p)^2}{2\sigma^2}\right]\left(0\le x_i \le L_S \text{ or } L_x - L_D \le x_i \le L_x\right) \\[2em] \dfrac{N_p}{\sqrt{2\pi}\sigma}\,\exp\left[-\dfrac{(x_i - L_s)^2 + (y_j - R_p)^2}{2\sigma^2}\right] + \dfrac{N_p}{\sqrt{2\pi}\sigma}\,\exp\left[-\dfrac{(x_i\,(L_x - L_D))^2 + (y_j - R_p)^2}{2\sigma^2}\right] \\[1em] \qquad\qquad (L_S \le x_i \le L_x - L_D) \end{cases}
$$

Fig. 3.22 Implanted Gaussian profile for the source/drains.

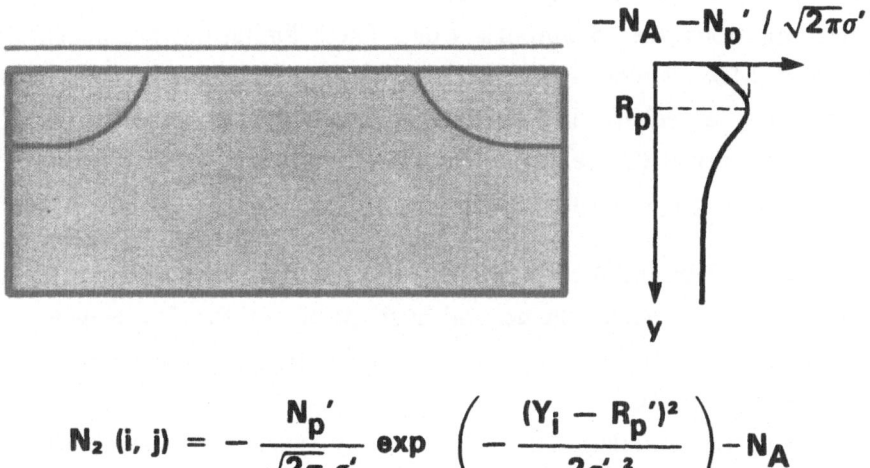

$$N_2\ (i,\ j)\ =\ -\ \frac{N_p{'}}{\sqrt{2\pi}\ \sigma'}\ exp\ \left(-\ \frac{(Y_i - R_p{'})^2}{2\sigma'^{\ 2}}\right) - N_A$$

Fig. 3.23 Implanted Gaussian profile for the channel.

0*	VGS	VDS	VBG	ID
	0.000	.100	0.000	3.666E-14
	0.500	.100	0.000	2.572E-09
	1.000	.100	0.000	8.693E-07
	1.500	.100	0.000	2.281E-06
	2.000	.100	0.000	3.592E-06
	2.500	.100	0.000	4.810E-06
	3.000	.100	0.000	5.950E-06
	3.500	.100	0.000	7.039E-06
	4.000	.100	0.000	8.079E-06
	4.500	.100	0.000	9.080E-06
	5.000	.100	0.000	1.005E-05

Fig. 3.24 Summary file of CADDET.

3.3 SIFCOD : General-Shape 2-D 2-Carrier Device Simulator

As mentioned in the previous sections, GEMINI and CADDET have
some limitations due to simplifications. These limitations are acceptable
for most MOS device applications, however, they are fatal in other
applications, especially in the hot electron related problem, CMOS
latch-up or novel device structures. The requirements for the ultimate
2-D device simulator are listed below.

1) General shape input : To simulate the practical device structure, the
   input structure must   be general in shape; if not, the device geometry
   should be simplified. This can become a major source of error in
   some applications.

2) Two carrier simulation : To completely specify the semiconductor
   device operation, both holes and electrons should be simulated. If not,
   some phenomena such as substrate current, latch-up, break-down due
   to the parasitic bipolar, and ionizing radiation effect cannot be
   simulated.

3) Transient simulation coupled with the passive circuit elements : The
   transient behavior of MOS transistors has been modeled using a DC
   current source and the intrinsic non-linear capacitance based on the
   quasi-static assumption. Though the quasi-static assumption is still
   valid even in the VLSI circuit considering the circuit speed and the
   intrinsic channel transit time, the intrinsic capacitances are difficult
   to model, especially in the short-channel device due to the 2-D field
   coupling and the velocity saturation.   Thus, direct 2-D transient
   device simulation coupled with the circuit elements is essential for
   predicting the performance of the new high-speed or scaled devices in
   the circuit. An alternative to direct performance simulation is the use
   of quasi-static models for drive current, and device and circuit
   parasitic models. The quasi-static approach is adequate for most
   digital circuits, but will not give good small-signal results for
   frequency comparable to maximum frequency of power-gain.
   Futhermore, it is impossible to model many transient phenomena in
   digital circuits such as the latch-up or ionizing radiation with the

static model.

In 1983, Mock from University of Jerusalem developed a new program, SIFCOD (SImulator For COupled Devices), to satisfy the needs of 2-D semiconductor device simulation with reasonable calculation time. SIFCOD program was acquired by Hewlett Packard Laboratories in late 1983. The program is essentially a combination of a very simple circuit analysis program with a package which directly simulates the transient of the devices and extracts the equivalent circuit for the given time and bias.   Mock's device transient simulation is based on his new algorithm [3.15], which enables the iterative solution of the Poisson and two current continuity equations during a transient and thus reduces the simulation time significantly. It will be explained in the next section in detail.   It uses the standard 5-point finite-difference method in discretization and ICCG3 [3.16] is used as a matrix solution method. Several enhancements have been made to improve the user interface and accuracy of SIFCOD. They are

1) Input method of device structure and doping : In the original program, the device structure and doping is input in the user defined subroutine. Thus, the program must be compiled and loaded again for the new device structure. It is very inconvenient for the user and difficult for the system manager to maintain and support. SIFCOD has been modified to read the SUPRA output file for the device structure and doping profile.

2) New mobility models for the electron and hole : To improve the accuracy of the drain current, the same electron and hole mobility models as those of CADDET have been implemented.

3) DC device simulation : Although steady-state device simulations are embedded in SIFCOD as an initial solution for the transient, users have neither control of the bias step nor of the program output of the terminal currents. Nonetheless, there are tremendous advantages if the steady-state and transient can be simulated by the same program, because each device can be bench-marked with measurements or examined in the steady-state simulation before the

time consuming transient simulation. Thus, the initial solution part has been modified to simulate the steady-state case.

4) Graphical post processing : SIFCOD has been modified to save the internal data as well as the terminal voltages and current so that they can be graphically analyzed by the post-processing program.

With all these enhancements, SIFCOD can accept any device structure which can be generated by SUPRA. Then, its steady-state characteristics can be simulated and examined to check its accuracy. In this device simulation step, the internal distribution can be checked by the graphical post processor. After that, the passive circuit elements are connected and the circuit performance is simulated using the transient part.

Basic Equations

The operation of the semiconductor device can be completely specified by both Poisson's equation and the time dependent electron and hole continuity equations with proper boundary conditions. The basic equations are

$$\nabla \cdot (\varepsilon \nabla \psi) = -q(p - n + N_D^+ - N_A^-) \tag{3.16a}$$

$$\frac{\partial n}{\partial t} = \frac{1}{q} \nabla \cdot \mathbf{J}_n - R \tag{3.16b}$$

$$\frac{\partial p}{\partial t} = -\frac{1}{q} \nabla \cdot \mathbf{J}_p - R \tag{3.16c}$$

$$\mathbf{J}_n = q\mu_n[\frac{kT}{q}\nabla n - n\nabla \psi - n\nabla \log(n_{ie}/n_i)] \tag{3.16d}$$

$$\mathbf{J}_p = -q\mu_p[\frac{kT}{q}\nabla p + p\nabla \psi - p\nabla \log(n_{ie}/n_i)] \tag{3.16e}$$

where ψ is the electrostatic potential; n and p are the electron and hole densities; ε is the dielectric constant; N_D^+ and N_A^- are the ionized donor and acceptor impurity densities; μ_n and μ_p are the electron and hole mobilities; n_i is the intrinsic carrier density and n_{ie} is the effective carrier density in heavily doped crystal. J_n and J_p are the electron and hole current densities.   The third term in the electron and hole current transport equations represent the current flow due to the change of n_{ie}.

For the transient simulation, these three   equations with the time-dependent terms should be solved. When they are solved iteratively as in the Gummel algorithm of the steady state, the time step should be smaller than the dielectric relaxation time[3.17]. This is a severe limitation in the MOS circuit simulation because the dielectric relaxation time is in the order of 10^{-14} seconds whereas the typical switching time in MOS circuit is in the order of 10^{-9} seconds. One way to eliminate the time step limitation is to solve the three equations simultaneously. This method,   however, increases the number of variables by three times. When the number of grid points is 2500, for example, the number of linear equations to be solved is 7500. It increases the computation time drastically.   In SIFCOD,   new equations which are equivalent to Eqs. (3.16) are employed. This method was proposed by Mock[3.15] and enables the iterative solution without a time step limitation. They are

$$\nabla \bullet (\varepsilon \nabla \frac{\partial \psi}{\partial t}) = - \nabla[(\mu_n n + \mu_p p)\nabla \psi] + \nabla[\mu_n \nabla n - \mu_p \nabla p] \quad (3.17a)$$

$$\frac{\partial n}{\partial t} = \frac{1}{q}\nabla \bullet \mathbf{J}_n - R \quad (3.17b)$$

$$\frac{\partial p}{\partial t} = - \frac{1}{q}\nabla \bullet \mathbf{J}_p - R \quad (3.17c)$$

In this formulation, the Poisson equation is differentiated with respect to time and the time derivatives of the electron and hole are replaced using the electron and hole continuity equations. Both continuity equations remain   the same as before.   These basic equations need additional auxiliary equations for the electron and hole mobilities, the net recombination, and the effective intrinsic carrier density which represents the bandgap narrowing in the high doping concentrations.

The same electron mobility model used in CADDET has been implemented for the electrons.   Using the same methodology, hole mobility for zero parallel field has been measured from a large number of p-channel devices and implemented in SIFCOD. The modified Gummel's mobility model for the surface channel is

$$\mu(E_n, E_p) = \frac{\mu_o(E_n)}{[1 + \frac{\{\mu_o(E_n)E_p/V_c\}^2}{\mu_o(E_n)E_p/V_c + G} + \{\frac{\mu_o(E_n)E_p}{V_s}\}^2]^{1/2}} \tag{3.18}$$

where E_p is the field parallel to the current flow and E_n is the field normal to the current flow. For electrons, the E_n dependence and coefficients are

$$\begin{aligned} \mu_o(E_n) &= 690 \ (E_n/10^5)^{-0.28} \\ V_s &= 1.036 \times 10^7 \quad cm/\ sec \\ V_c &= 4.9 \times 10^6 \quad cm/\ sec \\ G &= 8.8 \end{aligned} \tag{3.19}$$

For holes, the E_n dependence and coefficients are

$$\begin{aligned} \mu_o(E_n) &= (828.56 - 55.87 \log E_n) \\ V_s &= 1.2 \times 10^7 \quad cm/\ sec \\ V_c &= 2.128 \times 10^6 \quad cm/\ sec \\ G &= 1.6 \end{aligned} \tag{3.20}$$

For the effective intrinsic carrier density, Slotboom's model [3.18] [3.19] has been used.

$$\log \ (\frac{n_{ie}}{n}) = \frac{qV_o}{2kT}[\xi + (\xi^2 + \frac{1}{2})^{1/2}] \tag{3.21a}$$

$$\xi = \log \ (\frac{N_A + N_D}{N_O}) \tag{3.21b}$$

where V_O is 9 mV and N_O is 1.0×10^{17} cm^{-3}.

In the recombination, both Hall-Schockley-Read and Auger components are included and given by

$$R = (np - n_{ie}^2)[C_n n + C_p p + \frac{1}{\tau_p(n + n_{ie}) + \tau_n(p + n_{ie})}] \tag{3.22}$$

where C_n is 2.8×10^{-31} and C_p is 9.9×10^{-32} in cgs unit.

Program Organization and Numerical Algorithm

This section will explain how the program gets the input, discretizes the basic equations, solves the matrix equations and finds the terminal characteristics. The flow chart of the program is illustrated in Fig. 3.25. When the program is started, it first initializes the variables, reads the parameters from the ASCII input file and echoes the input to the

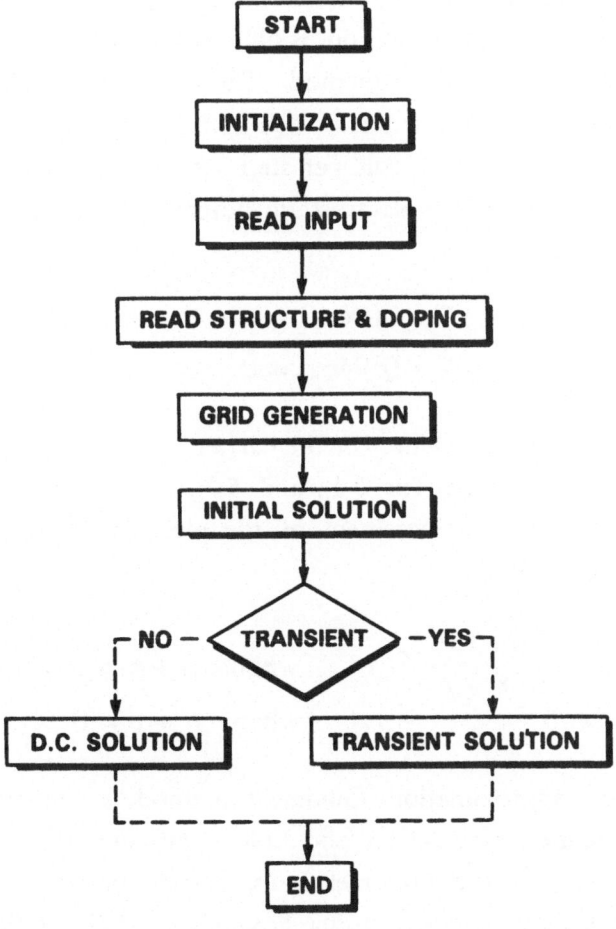

Fig. 3.25 Flow-chart of SIFCOD program.

terminal. The device structure and impurity distribution are, then, read from the binary SUPRA output file. Based on the impurity distribution, the x and y grids are automatically allocated. This easiness of the automatic grid allocation is one of the important advantages of the finite-difference method. After the initial preparation of the various device models, the initial solution for each device is computed with the initial bias voltages given by input. Based on this initial solution, the steady-state or transient solutions are obtained according to the input. Fig. 3.26 illustrates the flow chart of the initial solution. In the initial solution, an equilibrium solution is first obtained with all contact voltages set to zero, except the gate voltage in the MOSFET, which is set to the initial bias voltage. This initial solution is obtained by the solution of the Poisson equation using Newton's method. The device bias voltages are increased to their initial bias in steps, with a maximum change of any contact voltage limited to half volt per step. At the beginning of each step, an approximation of the electrostatic potential is computed using the following equations

$$\nabla \cdot (n\nabla\delta\phi_n) = 0 \qquad\qquad (3.23a)$$

$$\nabla \cdot (p\nabla\delta\phi_p) = 0 \qquad\qquad (3.23b)$$

using the presently available values for the carrier densities. On the ohmic contacts, the boundary values for $\delta\phi_n$ and $\delta\phi_p$ are the increments of the bias voltages. Then the increment of the electrostatic potential is obtained at each point from

$$\begin{aligned} \delta\psi &= \frac{n\delta\phi_n + p\delta\phi_p}{n + p} \qquad \text{when} \quad n + p > 0 \\ &= 0 \qquad\qquad\qquad \text{when} \quad n + p = 0 \end{aligned} \qquad (3.24)$$

Based on this approximation, Gummel's method is applied to the solution of the stationary electron and hole continuity equations and Poisson's equation. In the Gummel loop, carrier mobility and the recombination are calculated first using Eqs. (3.18) ~ (3.22). Using these parameters, electron and hole continuity equations in Eqs. (3.16) without

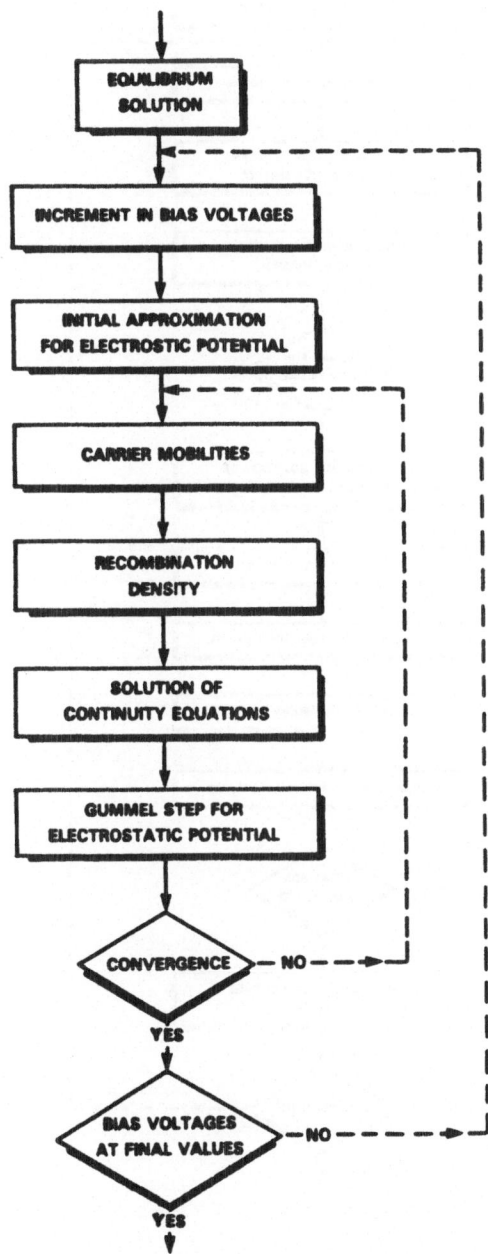

Fig. 3.26 Flow chart of steady-state solutions.

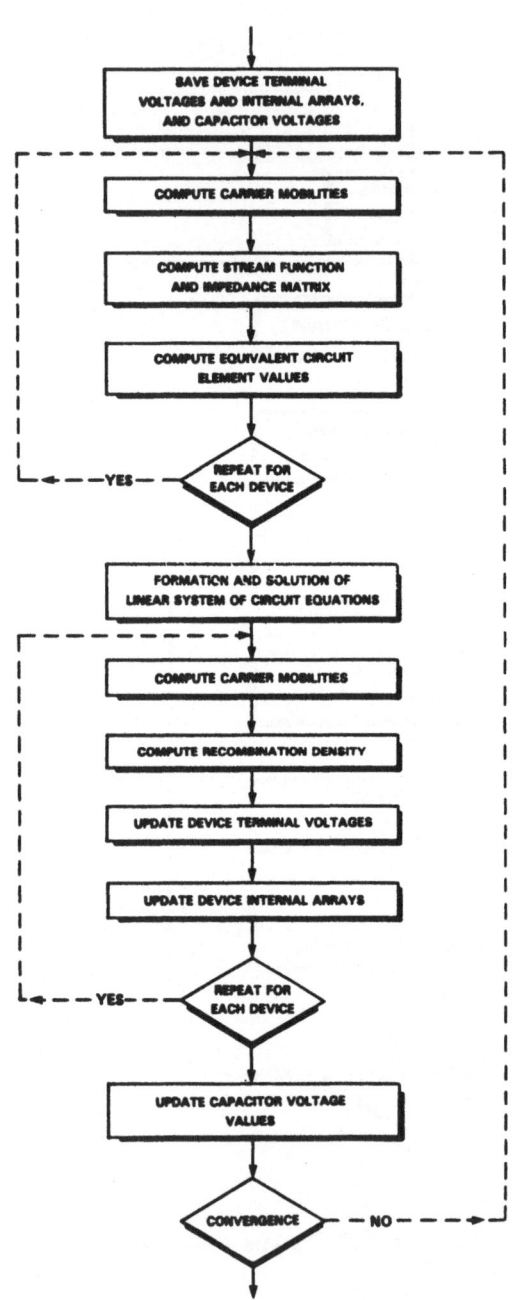

Fig. 3.27 Flow chart of the transient solutions.

the time-dependent term are solved with the fixed electrostatic potential. After that, the Poisson equation is solved for the electrostatic potential with the given electron and hole densities. This procedure is repeated until the potential converges. After convergence of the Gummel loop, bias voltages are stepped and the same procedure is repeated until the device reaches the initial bias condition.

In the steady-state solution, the same procedure as in the initial solution is repeated except that terminal currents are calculated after the convergence of the Gummel loop.

For the transient solution, the new basic equations in Eqs. (3.17) should be solved. The flow chart of the transient simulation is illustrated in Fig. 3.27. Based on the previous solutions, the equivalent circuit of each device is calculated. Combining these equivalent circuits with other circuit elements, the linear system of the circuit is formulated and solved for the node voltages. This will set the new bias voltages for the active devices. For each device, the new mobility and the recombination are calculated and the device terminal voltages are updated. The internal variables (potential, electron and hole density) are solved and updated using Eqs. (3.17). Finally, the capacitor voltages are updated and the same procedures are repeated until convergence is obtained. At each time step, the device terminal voltages, internal arrays(ψ, n, p) and capacitor voltages of the previous time step are saved.

Example

The following example is the DC simulation of an enhancement-mode MOSFET. The gate oxide thickness of the device is 40 nm and the effective channel is 1.25 μm. Since SIFCOD accepts the input device geometry and doping profile from SUPRA output, the SUPRA input of the device is listed in Fig. 3.28. The resulting device structure and impurity distributions are plotted in Fig. 3.29. The plot shows that the junction depth of the source/drain is 0.35 μm and the lateral diffusion is 0.3 μm. The SUPRA output is saved in a binary file for SIFCOD. The input of SIFCOD is listed in Fig. 3.30. Since SIFCOD uses the namelist

```
TITLE       Enhancement MOS device : STEP 1

STRUCTURE P-TYPE ORIEN=100 DEPTH=3.0 WIDTH=1.5 HEIGHT=0.5
+           CONCEN=4E14
X.GRID      H1=0.05  H2=0.05   WIDTH=1.50 N.SPACES=30
Y.GRID      H1=0.20  H2=0.05   DEPTH=0.47 N.SPACES=4
Y.GRID      H1=.002  H2=.002   DEPTH=0.02 N.SPACES=10
Y.GRID      H1=.004  H2=.004   DEPTH=0.02 N.SPACES=5
Y.GRID      H1=.002  H2=.002   DEPTH=0.02 N.SPACES=10
Y.GRID      H1=.004  H2=.20    DEPTH=2.97 N.SPACES=30
END         STRUCTURE DEFINITION

COMMENT     START OUT ANALYTIC MODE
ANALYTIC
COMMENT     GATE OXIDE DEPOSITION
OXIDIZE     TEMP=850 TIME=33.0 WET
OXIDIZE     TEMP=850 TIME=19.0 INERT

COMMENT     CHANNEL IMPLANT
IMPLANT     BORON DOSE=6E11 ENERGY=50

COMMENT     POLY DEPOSITION
DEPOSIT     POLYSILICON THICKNESS=0.4

COMMENT     DEFINE POLY
ETCH        POLYSILICON START=0.55 END=0

COMMENT     POLY DOPING
OXIDIZE     TEMP=900 TIME=65 INERT

COMMENT     SWITCH TO NUMERIC
NUMERICAL
COMMENT     ARSENIC S/D IMPLANT
IMPLANT     ARSENIC DOSE=7E15 ENERGY=90

COMMENT     S/D DRIVE IN
OXIDIZE     TEMP=850 TIME=65 INERT

COMMENT     DEPOSIT CAP OXIDE
COMMENT     S/D JUNCTION DRIVE-IN
OXIDIZE     TEMP=950 TIME=45 INERT

COMMENT     SAVE STRUCTURE FOR PLOT AND GEMINI
SAVE        STRUCTURE=ENSD1
END

COMMENT     Formation of GEMINI structure file
COMMENT     Create the total transistor structure
STRUCTURE
LOAD        STRUCTURE=ENSD1.DAT X.MAX=1.45
LOAD        STRUCTURE=ENSD1.DAT X.MAX=1.45 REFLECT
END         Structure definition

COMMENT     Save the GEMINI structure file
SAVE        STRUCTURE=SUPBIN2.DAT  GEMINI
END
```

Fig. 3.28 SUPRA input of the example device.

feature of FORTRAN in its input, all input cards start with '&'+ keyword and end with &END. Between these two delimiters, parameters are specified with the parameter name and value separated by equal signs. The parameters are separated by commas. The first card should be &DIM card. It specifies the maximum dimension of the major arrays. KM is the largest number of the horizontal grids and KN is the largest number of the vertical grids. Next, the &IN card specifies the circuit elements and its connections. It is also used to initiate the DC and transient solutions. In a &IN card, parameter I designates the type of the circuit element or solution. When I is from 1 to 12, it specifies the type of the circuit element and connections. The circuit elements can be resistors, capacitors, voltage or current sources, MOSFETs, bipolar transistors, or general devices. For more details, refer to the SIFCOD manual [3.20].

In this example, &IN cards with I=3 is for the voltage sources. &IN card with I=12 specifies the general device. NUM is the number of contacts on the device. For the enhancement mode MOSFET device, the

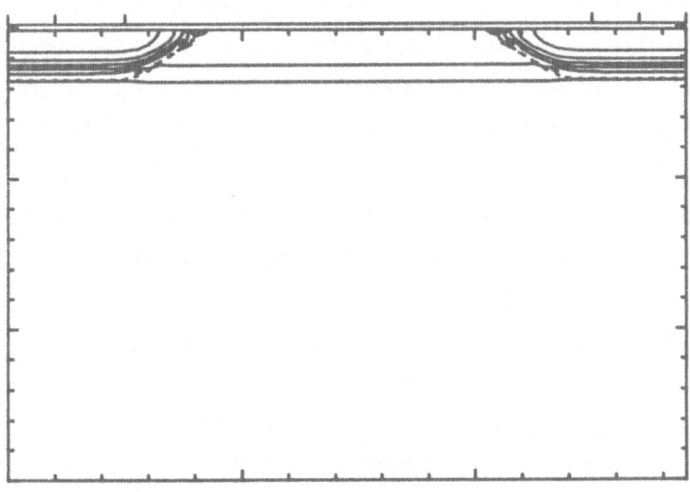

Fig. 3.29 SUPRA output of the example device.

number of contacts is four. MODEL specifies the model for this device. NODE is used to specify the circuit connection of the device contacts. BV is the initial bias voltages. The values of BV are given in the order of contacts. The device model is given in the next &IN card with I=13 and &DEV card. In the &IN card, for the device model, MODEL specifies the model identification number. OTGEM is the file name of the SUPRA output, from which the device geometry and doping profile are read. In the &DEV card, M is the number of the horizontal (x) grid lines and N is the number of the vertical (y) grid lines. In SIFCOD, the origin of the coordinate is the upper left corner of the simulation region. The positive direction of x is to the right side and the positive direction of y is to the bottom. IXZ and IYZ specify the grid allocation methods for x and y grids respectively. When IXZ or IYZ is 4, the grid is computed based on the doping profile along the line y=XZL or x=YZL. When IXZ or IYZ is 7, the SUPRA grid is used. DXMIN and DXMAX are the minimum and maximum spacing in x grid. DYMIN and DYMAX also specify the minimum and maximum spacing in y grid. In SIFCOD, the external contact can be located only along the sides of the rectangular simulation region. NSIDE selects the side on which the contact will be put. The top side is 1 and the bottom is 3. The right and left sides are 2 and 4 respectively. Z specifies the position of contacts on the selected sides by pair of coordinates. A contact starts from the first coordinate and ends at the second coordinate. Here, contact 1 is the gate on the top side and contact 3 is the substrate on the bottom. Contact 2 is the drain on the right side and contact 4 is the source on the left side. Q is the Q matrix in column-wise order. The Q matrix specifies the current components in the stream function calculation. For more detail, refer to Mock's book[3.15]. The last &IN card with I=14 initiates the DC solutions. Here, the bias voltages of two contacts can be changed in one card. NODEM identifies the primary node to be changed and NODEP is the secondary node. VMSTP and VPSTP are the voltage steps for the primary and secondary nodes respectively. NMSTP and NPSTP are the number of bias steps of primary and secondary nodes. In this example, the bias voltage of the secondary node (gate) increases by 1 volt from its initial bias and the

```
&DIM KM=30    ,KN=71    ,&END
&IN  I=3      ,NF=1     ,A=5.     ,&END
&IN  I=3      ,NF=2     ,A=5.     ,&END
&IN  I=3      ,NF=3     ,A=0.     ,&END
&IN  I=12     ,NUM=4    ,MODEL=1
     ,NODE=1,2,3,0
     ,BV=0.,0.,0.,0.,&END
&IN  I=13     ,MODEL=1 ,OTGEM='SUPRA.DAT',&END
&DEV M=30     ,N=71     ,IGEOM=1,IXZ=4  ,IYZ=4
     ,NSIDE=1,2,3,4
     ,XL=2.8 ,XD=3.0   ,XW=48.6,XZL=0.6,YZL=1.45
     ,DXMIN=0.05       ,DXMAX=0.10      ,DYMIN=0.0002
     ,Z=0.55,2.35,0.5,0.6,0.0,2.8,0.5,0.6
     ,Q=1.,0.,-1.,0.,-1,1.,0.,0.,0.,1.,0.,-1.,&END
&IN  I=14     ,NUM=1
     ,NODEM=2,VMSTP=0.5,NMSTP=10
     ,NODEP=1,VPSTP=1.0,NPSTP=5
     ,&END
```

Fig. 3.30 Input of SIFCOD for NMOS example.

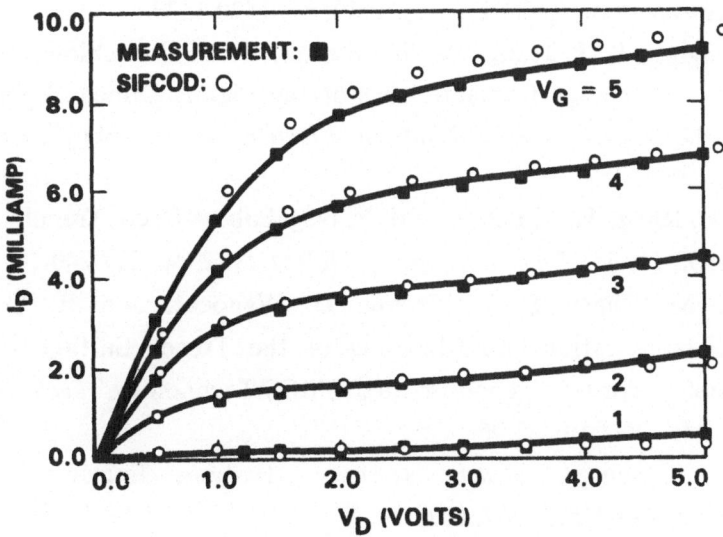

Fig. 3.31 Comparison of the simulations with measurements.

bias voltage of the primary node (drain) changes from 0 to 5 volts by 0.5 volts step. The same procedures are repeated for each increase in gate voltage until it reaches 5 volts.

After preparing the SUPRA output and SIFCOD input, SIFCOD can be run in batch mode. SIFCOD has two output files; one is for the summary of bias and terminal currents and the other is a data file for graphical plotting. In the batch command, file names of the input, summary output and plot output should be given. SIFCOD first reads the input file and echoes it. After that, it starts to solve the initial solution by iterating Poisson and two current continuity equations. Then it steps the bias voltages and iterates for the steady state solutions. The drain currents are plotted and compared with the measurements in Fig. 3.31. Agreement with measurements is within 10%.

References

[3.1] J. A. Greenfield and R. W. Dutton, "Nonplanar VLSI Device Analysis Using the Solution of Poisson's Equation," *IEEE Trans. on Electron Devices*, ED-27, Aug 1980, pp. 1520-1532.

[3.2] S. Ogura, P. J. Tsang, W. W. Walker, D. L. Chritchlow, and J. F. Shepard, "Design Characteristics of the Lightly Doped Drain-Source (LDD) IGFET," *IEEE Trans. on Electron Devices*, ED-27, Aug 1980, pp. 1359-1367.

[3.3] R. D. Rung, H. Momose, and Y. Nagakubo, "Deep Trench Isolated CMOS Devices," *Tech. Digest of IEDM 1982*, pp. 237-240.

[3.4] K. M. Cham, S. Y. Chiang, D. Wenocur, and R. D. Rung, "Characterization and Modeling of the Trench Surface Inversion Problem for the Trench Isolated CMOS Technology," *Tech. Digest of IEDM 1983*, pp. 23-26.

[3.5] R. S. Verga, *Matrix Iterative Analysis*, Englewood Cliffs, NJ:Prentice-Hall, 1962, ch.6.

[3.6] J.M. Ortega and W. C. Rheinboldt, *Iterative Solution of Nonlinear Equation in Several Variables*, New York:Academic Press, 1970, pp. 214-230.

[3.7] T. Toyabe, *"CADDET User's Manual"*, Hitachi Central Laboratories

[3.8] M. S. Mock,*"Analysis of Mathematical Models of Semiconductor Devices,"* Boole Press, Dublin, 1983.

[3.9] H. L. Stone, "Iterative Solution of Implicit Approximations of Multidimensional Partial Difference Equations," *SIAM J. Numerical Anal.*, 5, 1968, pp. 530-558.

[3.10] D. L. Scharfetter and H. K. Gummel, "Large-signal Analysis of a Silicon Reed Diode Oscillator," *IEEE Trans. on Electron Devices*, ED-16, pp. 64-77, Jan 1969.

[3.11] K. Yamaguchi, " Field-Dependent Mobility Model for Two-Dimensional Numerical Analysis of MOSFETs," *IEEE Trans on Electron Devices*, Vol ED-26, pp. 1068-1074, July 1978.

[3.12] K. Yamaguchi, " A Mobility Model for Carriers in the MOS Inversion Layers," *IEEE Trans. on Electron Devices*, Vol Ed-30, pp. 658-663, June 1983.

[3.13] S. Y. Oh, P. Vande Voorde, and J. Moll, "An Empirical Mobility Model for Numerical MOSFET Simulation," *Hewlett-Packard Semiconductor Technology Conference Proc.*, pp. 97-104, 1984.

[3.14] K. K. Thornber, "Relation of Drift Velocity to Low-Field Mobility and High Field Saturation Velocity," *J. Appl. Phys.*, Vol. 51, No. 4, pp. 2127-2136, April 1980.

[3.15] M. S. Mock, *"Analysis of Mathematical Models of Semiconductor Devices ,"* Boole Press, Dublin, 1983.

[3.16] J. A. Meijerrink et al, "An Iterative Solution Method for Linear System of Which the Coefficient Matrix is a Symmetric M-Matrix," *Mathematics of Computation*, Vol 31, No 137, Jan 1977, pp. 148-162.

[3.17] J. J. Barnes, *"A Two-dimensional Simulation of MESFETs,"* Ph.D dissertation, University of Michigan, Ann Arbor, Aug 1975.

[3.18] J. W. Slotboom et al, *Solid-State Electronics*, 15, 1972, pp. 1229-1235.

[3.19] A. Wieder, *IEEE Trans. Electron Devices*, ED-27, 1980, pp. 560-607.

[3.20] M. S. Mock, *"The SIFCOD program User's Guide,"* June 1983.

Chapter 4

FCAP2 : Parasitic Capacitance/Resistance Simulator

4.1 Introduction

To simulate the parasitic capacitance/resistance, it is necessary to solve the Poisson (or Laplace) equation in at least two dimensions with arbitrary input geometry. FCAP2 is the dedicated 2-D program for this purpose. The basic features of the program are;

1) It solves the linear Poisson equation using the standard 5-point finite-difference method to make the automatic grid generation easy.

2) The input geometry can be arbitrarily described by the combination of the basic elements such as rectangle, triangle, circle and arc. It is very useful to simulate the cross-section of the real process topology.

3) Reflective boundary conditions are used to reduce the computation in the symmetric or repetitive structure. In the open boundary (boundary goes to infinity), however, care should be taken to put the boundary far away from the region of interest to reduce error due to the reflective boundary conditions.

4) The grid generation is automatic and transparent to the user, which makes the program easy to use. To enhance accuracy, a variable grid is used and the grid can be remapped according to the previous solution.

5) ICCG [4.1] (Incomplete Cholaski Conjugate Gradient method) is employed in matrix solution, which is one of the fastest methods. It

< RECTANGLE >

RECT VER1 = a,b VER2 = c,d [ROT = r] [CHRG = q]   VOLT = V
                                                    DIEL = d   NAME = nm

< TRIANGLE >

UTRI
LTRI   VER1 =   VER2 =   [ROT =] [CHRG =]   VOLT =   NAME =
                                                 DIEL =

< CIRCLE & ARC >

CIRC ORG = a,b RADV = c,d [ANGL = r] [CHRG = q]   VOLT =   NAME = nm
                                                   DIEL =

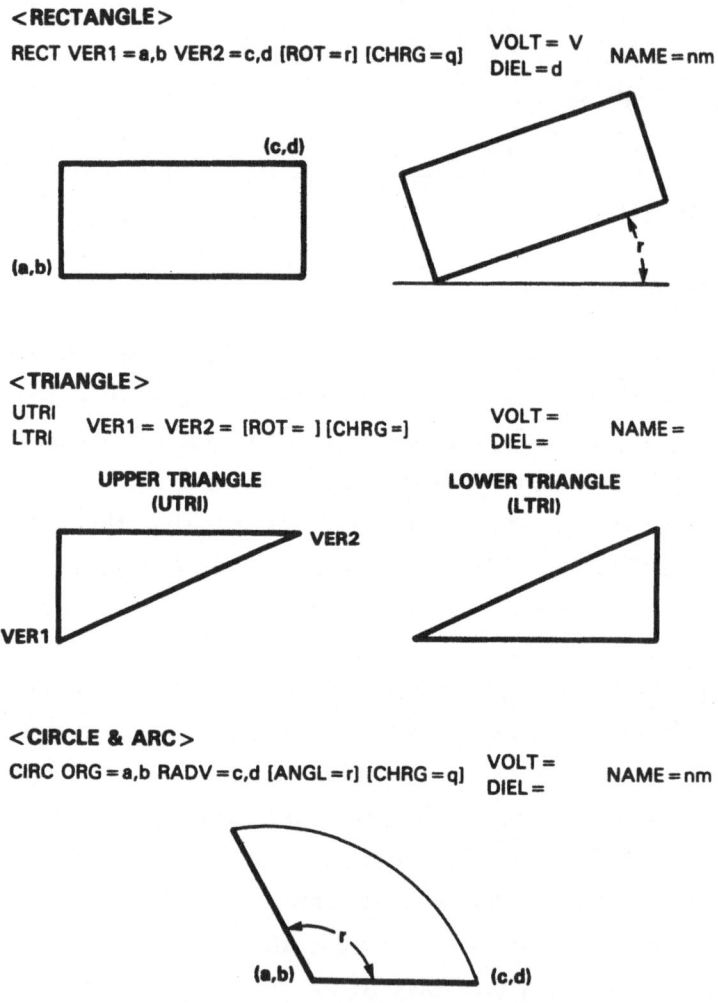

Fig. 4.1 Basic input elements of FCAP2.

speeds up the computation by several orders of magnitude compared to other matrix solution methods.

With all these features, FCAP2 is the ideal tool to simulate the capacitances of IC and microstrip lines. Its applications, however, are not limited in those cases. It is a general 2-D Poisson solver so that all the phenomena specified by the Poisson or Laplace equations can be simulated such as

1) Potential and field distributions in the electro- and magneto-static problems
2) Resistance calculations in resistive materials
3) Stationary thermal problems.

4.2 Input and Flowchart of FCAP2

In FCAP2, the arbitrary input geometry is specified using the basic geometry elements. They are rectangle, upper triangle, lower triangle, arc and circle. The input geometry file which describes the geometry to be simulated should be prepared before running the program. The format of the input geometry commands is illustrated in Fig. 4.1. All the commands start with an element name. The parameters of each element are specified by the parameter key word and its values. The unit of length is in microns, the rotation in degrees, charge per cubic centimeter and the voltage in volts. The dielectric constant is the relative dielectric constant. In the rectangle, upper and lower triangles, VER1 and VER2 specify the location of the elements as shown in Fig. 4.1. The parameter ROT specifies the rotation of the elements from their original position. It is optional and its default value is 0. For the circle, the position is specified by ORG and RADV as shown in Fig 4.1 and ANGLE specifies the arc angle. The default value is 360 degrees. In all the elements, VOLT should be specified for the conductor and DIEL should be given for the dielectric. In the dielectric, CHARG is the optional parameter which specifies the density of the uniform charge distribution. The parameter NAME should be given for each elements which will be used to identify each element. At the end of the input, the commands BACK and BOX specify the material and the size of the background space in simulation. The END card should be placed at the end of the input. After the input geometry file is prepared, FCAP2 is run in interactive mode. The flowchart is shown in Fig. 4.2. Once the program starts, after reading the input file it asks for, it then plots the input geometry. If the geometry is right, the solution of 2-D Poisson equation can begin. When the solution converges, the potential or field distribution can be plotted and the charge on all the conductors can be calculated for capacitance. If better accuracy is desired, the grid can be remapped based on the calculated potential and the problem can be solved again using the new grid. This procedure can be repeated until the solution converges. The solution can then be saved for a better plot package and the program ends.

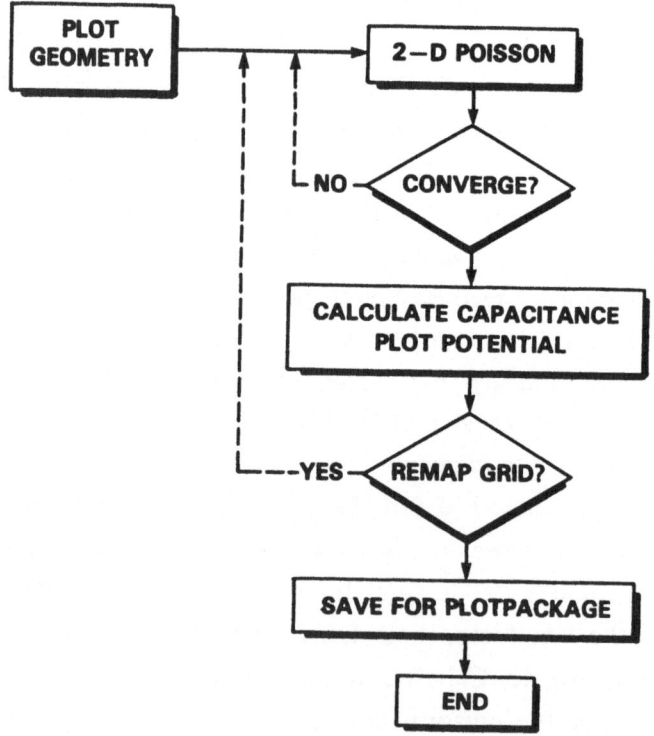

Fig. 4.2 Flow chart of FCAP2 program.

To give a more realistic feeling about how to run FCAP2, a simple example of FCAP2 run will be illustrated in the next section.

4.3 Example

In order to illustrate how FCAP2 works, a simple example of wiring capacitance simulation shall be explained. The input geometry is shown in Fig. 4.3(a). A metal wire is running over the silicon substrate. Its width is 4 microns and the thickness is 1 micron. The oxide thickness between the metal and the substrate is 0.5 micron. One micron thick oxide has been deposited on the poly wire. The input file is also shown in Fig. 4.3(b).

When FCAP2 runs, it prompts for the input file name first. After reading the input file, the program echoes the input. Next, it prompts for the plotting device and asks whether to plot grid points or not. It also prompts the plotting ranges and plots the input geometry. Type anything if you want to know what are then available commands at that stage. At this point, the only valid commands are the GEometry command for new input geometry, the ITerate command to solve the Poisson equation and the STop command to exit the program. The first two characters of each command are enough to execute it. Next, iterate to get the solution. During iteration, the convergence criterion and maximum number of iterations are requested. Usually, default values are enough. The program then iterates until convergence occurs. In each iteration, the program prints out the maximum residue (R(MAX)) and the tolerance of the iteration (EPS).

After convergence, the available commands are CHarge to calculate the charge per unit length for each conductor, COntour to plot the 2-D contour map, REmap to rearrange the grid for better accuracy based on the previous potential distribution and SAve to save the plot data for the plot package. GEometry, ITerate and STop commands are also available. Next, calculate the charge to evaluate the capacitance and plot the contour map. The charge is expressed in pico-coulomb per meter. For contour plotting, the options are POtential, EField and REsidue. Here, the

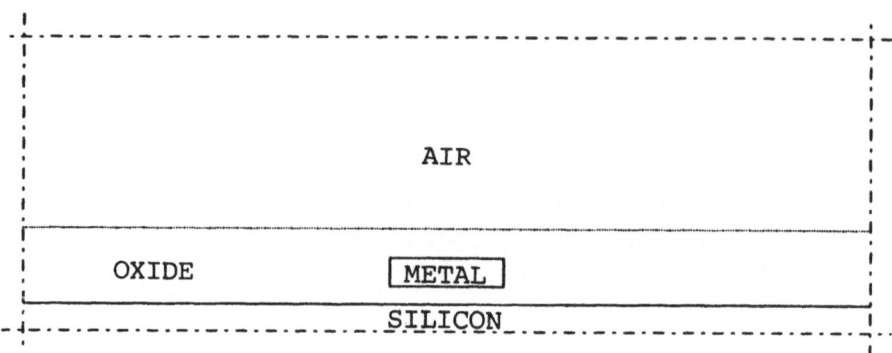

(a)

```
0001  RECT  VER1=0,0      VER2=30,3.5  DIEL=3.9  NAME=OX
0002  RECT  VER1=0,0      VER2=30,1    VOLT=0    NAME=SI
0003  RECT  VER1=13,1.5   VER2=17,2.5  VOLT=1    NAME=MT
0004  BACK  DIEL=1        NAME=BK
0005  BOX   XBOX=30       YBOX=10
0006  END
```

(b)

Fig. 4.3 (a) Input structure of the example (b) Input
file for the example.

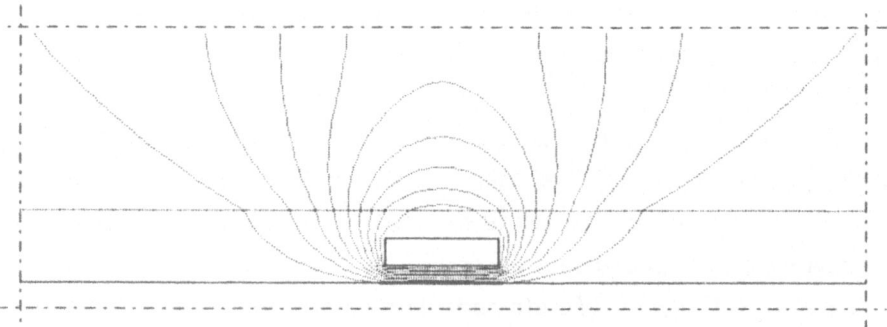

Fig. 4.4 Contour plot of potential distribution. The
value of the equi-potential line starts from 0.0
volt to 1.0 volt by 0.1 volts step.

user is prompted for the minimum, maximum and the increment of the contour. Fig. 4.4 shows the contour plot.

To enhance accuracy, remap the grid and iterate again. When it converges, calculate the charge and stop. This example shows that the charge can be changed more than 30% with the grid. To ensure accuracy, the remap and iteration are repeated until the charge calculation is stable.

Reference

[4.1] D. S. Kershaw, "The Incomplete Cholesky-Conjugate Gradient Method for the Iterative Solution of Systems of Linear Equations," Journal of Computation Physics 26, 1978, pp. 43-65.

Part B

Applications and Case Studies

Chapter 5

Methodology in Computer-Aided Design for Process and Device Development

The previous chapters have presented an overview of computer-aided design (CAD) in VLSI development, as well as the simulation tools currently used at Hewlett-Packard Laboratories. In this chapter, CAD is discussed from the user point of view. The methodology for using the simulation tools in the most effective way is presented. Then case studies will be presented in the following chapters which show in detail how simulation tools are used in device designs.

5.1 Methodologies in Device Simulations

Simulation tools should be used in the most efficient way, such that time and effort in doing the simulations are minimized. This is especially true in process development where time is of major concern. The goal is to provide process parameters in the shortest time. Ideally, one would like to have CAD tools which can produce the desired output in minimal time. In real life, this is usually not possible, mostly because of limitations in software and hardware capabilities, or due to too many users on the system. In the following paragraphs, we discuss the simulation methodologies which will use the CAD tools most effectively:

1) Before simulations are to be performed, it is always a good idea to look at the problem from the simplest and most intuitive point of view, and to try to get some basic idea of the problem. For example,

in the case of counter-doping for the p-channel transistors with n+
polysilicon gate (to be discussed in detail in Chapter 8), the dose can
be roughly estimated by simple arguments, and it turns out to be not
very different from the results of two-dimensional analysis. This
simple procedure not only provides a first estimate of the magnitude
of the process parameters to be used, but also provides a simple
picture of the physics behind the technique. Numerical analysis is
sometimes difficult to interpret unless one has a basic idea of the
physics involved. Also by gaining some initial knowledge of the
problem, the simulation work can be better planned and executed,
with a minimum range of parameter values, instead of using trial and
error.

2) A simulation should be considered as an experiment itself. This
means that a systematic procedure should be used rather than
shooting for a particular number. For example, it may be desired to
develop a p-channel MOSFET with a particular threshold voltage,
say, -0.7 V, and the threshold voltage is dependent on the
counter-doping of boron [5.1]. In the simulation work, a factorial
experiment [5.2] should be set up with the boron implant dose and
energy as parameters. The simulated results are shown in Fig. 5.1.
This figure provides an overall picture of the dependence of the
threshold voltage on the two parameters. Also it provides an idea of
the reasonable range of the parameters that should be used in the
fabrication experiment. To be more complete, the n-well impurity
surface concentration should be included in the simulation. In this
case, the simulations become a 2^3 factorial experiment. The results,
when arranged in a 3-dimensional form as shown in Fig. 5.2, show
the dependence of the threshold voltage on the three parameters.
The threshold voltage change due to the change in a combination of
the parameters can be estimated quickly from the figure.

Several observations can be made from this simple example.
Fig. 5.3 shows the threshold voltage (V_T) as a function of the
counter-doping dose (D_C) over a wide range. The dependence is
nonlinear over the whole range, but can be approximated to be linear

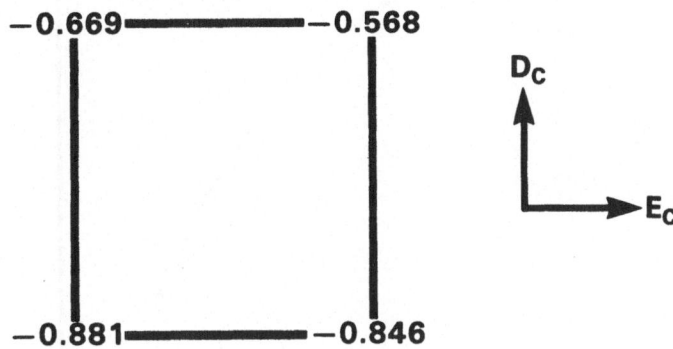

Fig. 5.1 P-channel threshold voltage vs. counter-doping
energy (E_C) and dose (D_C).

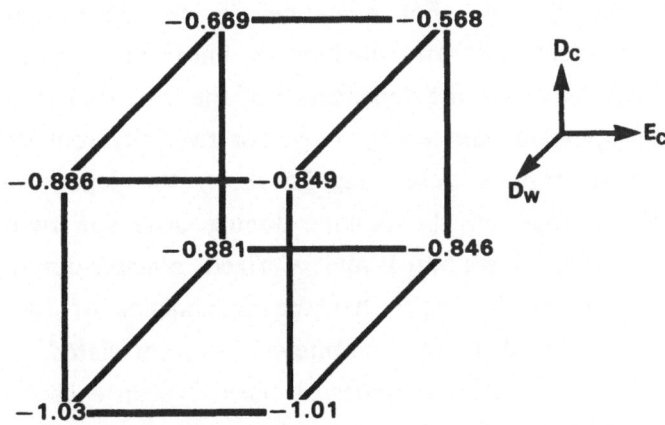

Fig. 5.2 P-channel threshold voltage vs. counter-doping
energy (E_C), dose (D_C) and n-well implant
dose (D_W).

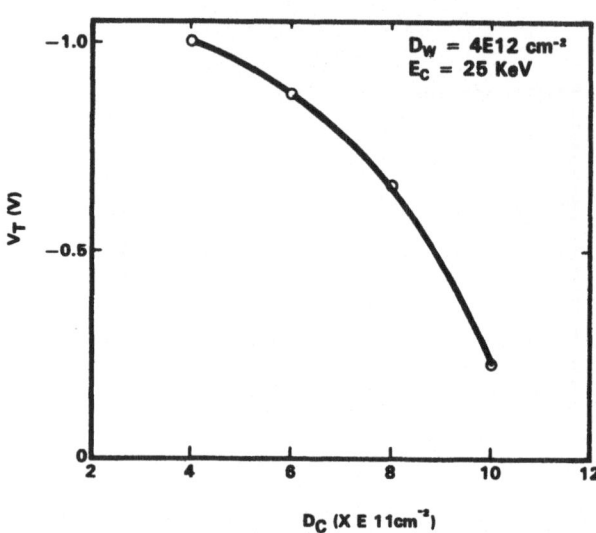

Fig. 5.3 P-channel threshold vs. implant dose D_C.

within a smaller range. (Nonlinearity is most serious for the counter-doping case. For n-channel device, the threshold versus channel implant dose are much more linear in a larger range of doses.) Fig. 5.4 shows the dependence of the threshold voltage on the counter-doping implant energy (E_C) for two different doses and a fixed n-well implant dose. Fig. 5.5 shows the dependence of the threshold voltage on the counter-doping dose for two different n-well implant doses (D_W) and a fixed counter-doping implant energy. The results show that the dependence of the threshold voltage on the different parameters are correlated, as can be observed by the different slopes of the curves in each figure. This means that for a narrow range of the parameters, the threshold voltage can be expressed as:

$$V_T = a_1 D_C + a_2 E_C + a_3 D_W \\ + b_{11} D_C E_C + b_{12} D_C D_W + b_{13} E_C D_W + c D_C E_C D_W \qquad (5.1)$$

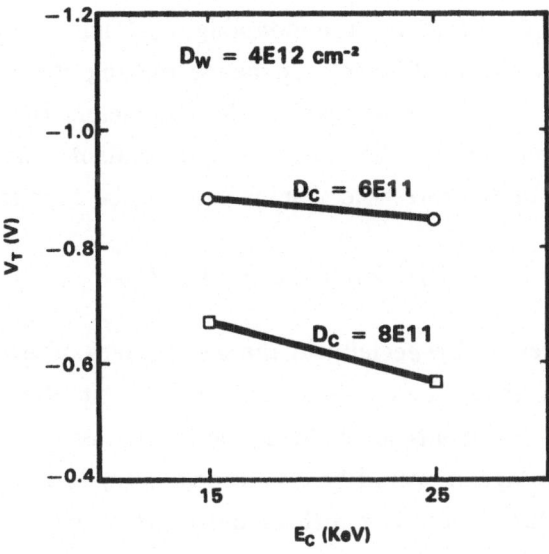

Fig. 5.4 SUPREM simulation of p-channel threshold vs. counter-doping energy for two different doses.

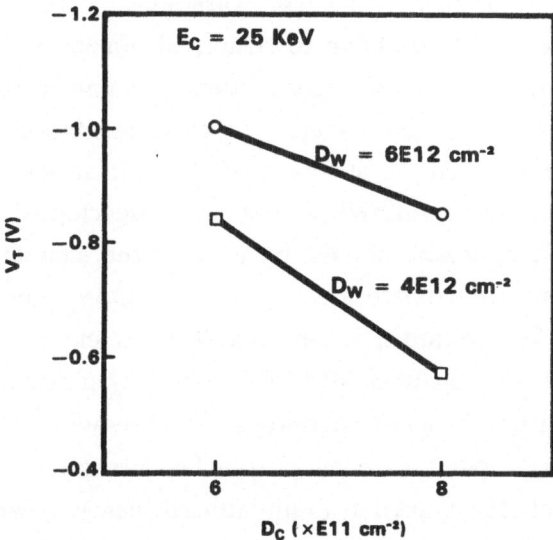

Fig. 5.5 SUPREM simulation of p-channel threshold voltage vs. counter-doping dose for two different n-well implant doses.

In most cases, however, the coupling coefficients b_{ij} and c are smaller than the coefficients a_i, hence making the problem much simpler. As a first approximation, the parameters D_C, E_C, and D_W can be assumed to be linearly independent variables. The dependence of the threshold voltage can thus be expressed as:

$$V_T \simeq a_1 D_C + a_2 E_C + a_3 D_W \qquad (5.2)$$

3) In most cases, and especially in the early stages of experiments, the trend rather than exact values generated by simulations should be emphasized. For example, in the above mentioned experiment on the threshold voltage, it would be a bad idea to try to generate the dose and energy that would give a threshold value of exactly -0.7 V. There are two reasons for this; first, the simulation tools are not always very accurate. The accuracy depends on the particular structure being considered and the simulation tools being used. Second, there are process control variations during fabrication, such as the line width and dielectric thickness variations. In VLSI, device performance is very sensitive to structural dimensions, so it is very important to understand the sensitivities. The simulations provide an optimized range of parameters for experimentation, which then provides feedback on the accuracy of the simulations. Through this feedback loop, the optimized process can be developed.

4) The simplest approach should be used in the simulations. It is a waste of time to simulate a structure with more accuracy than is necessary. For example, if one wants to calculate the threshold voltage of a long channel MOSFET, it is only necessary to use the SUPREM program, which considers the threshold voltage from a MOS capacitor point of view [5.3]. This method is valid because short channel effects such as drain-induced barrier lowering [5.4] are absent in this case. It would be a waste of time to use two-dimensional numerical programs in this case. There may be cases in which only rough estimates are needed to evaluate whether a particular idea is feasible. In such a case, the simplest procedures

that satisfy the relaxed accuracy requirements should be used. When the idea is considered feasible, and more detailed study is desired, then more sophisticated simulation tools will be used. Also, as mentioned in discussion 1), the simplest procedure usually provides the most basic and easily understood physical insight of the problem. In general, the idea is to choose the tool that provides the necessary accuracy, but not more.

5.2 Outline of the case studies

The following chapters are case studies in which simulations are used in process development. In Chapter 6, we discuss basic device physics for process engineers, and how to generate device parameters by using simulations. The relationship between device characteristics and process parameters are presented. Chapter 7 studies the drain-induced barrier lowering effect in short channel MOSFETs. Chapter 8 presents how CAD is used in submicron CMOS transistor design. The issues of subthreshold leakage and drain-induced barrier lowering, as well as other concerns in short channel MOSFETs are discussed. Chapter 9 presents the use of CAD tools in analyzing a special structure, which is the trench isolation in CMOS. This chapter shows how simulations have identified a potential problem, and provided recommendations on how to minimize such problem. Chapter 10 shows the use of simulation tools in the development of new isolation structures such as SWAMI. Simulations are used to evaluate the performance of isolation structures. In Chapter 11, a combination of simulations and experiments are used to study in detail the lightly-doped drain (LDD) structure. The physics of the structure are revealed through simulations. Chapter 12 shows how simulations can be used to optimize the scaling of a process to achieve higher circuit performance and minimize process complexity. Chapter 13 presents the simulation of parasitics in circuits. Parasitic resistances and capacitances are simulated which provide very important information essential for circuit simulations. The physics of parasitic capacitances is also discussed.

References:

[5.1] K. M. Cham and S. Y. Chiang, "Device Design for the Submicrometer P-Channel FET with n+ Polysilicon Gate," *IEEE Trans. on Electron Devices*, <u>ED-31</u>, July 1984, pp. 964-968.

[5.2] G. E. P. Box, W. G. Hunter, and J. S. Hunter, *Statistics for Experimenters*, NY:John Wiley & Sons, 1978.

[5.3] S. M. Sze, *Physics of Semiconductor Devices*, 2nd ed., NY:Wiley Interscience, 1981.

[5.4] R. R. Troutman, "VLSI Limitations from Drain-Induced Barrier Lowering," *Trans. on Electron Devices*, <u>ED-27</u>, April 1979, pp. 461-468.

Chapter 6

Basic Techniques in Simulations for Advanced Process Development

In this chapter, the basic simulation techniques for advanced MOS process development will be described. First of all, the basic device physics of MOSFET is presented. The discussion will be in very simple terms, although sufficient to allow the process engineers to understand the basic characteristics of MOSFETs and their significance. The techniques of generating the device parameters are then presented. Also to be discussed are the short channel effects such as drain-induced barrier lowering. Simulations are used to reveal details of these phenomena. The relationship between process parameters and device characteristics are discussed. Simulated results are compared with experimental results. The discussions will emphasize the GEMINI program since it is relatively simple to use.

6.1 Basic Device Physics For Process Development

The physics of the MOSFET has been discussed in detail in numerous books and papers [6.1]-[6.5]. The purpose of this section is not to reproduce those discussions, but rather to highlight in simple terms the critical parameters for process development, and discuss briefly their significance to the circuit performance.

The electrical characteristics of MOSFETs can be separated into two regions. One region is the subthreshold region, where the drain current of the transistor is exponentially dependent on the gate voltage. Fig. 6.1

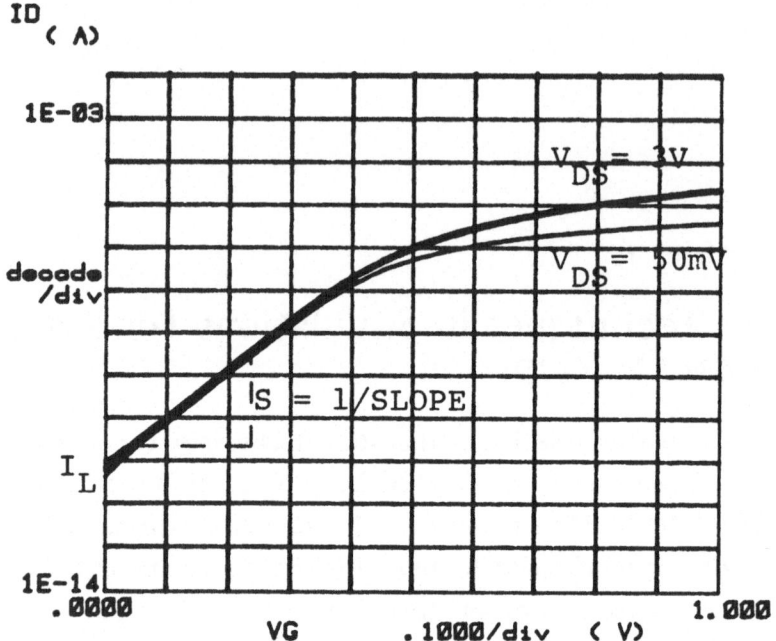

Fig. 6.1 Drain current (log. scale) vs. gate bias for a long channel MOSFET, with drain to source bias (V_{DS}) of 50 mV and 3 V.

shows the measured subthreshold characteristics of a long channel MOSFET. In this region, the most interesting parameters are the subthreshold slope, threshold voltage, and the residual current at zero gate bias. These parameters are also shown in the figure.

The subthreshold slope (S) is conventionally expressed as the inverse of the slope of the drain to source current (in logarithmic scale) versus the gate to source bias. S measures how fast the drain current can be turned off as a function of gate bias. It is typically between 80 to 120 mV/decade. The steeper the slope (hence smaller S), the less gate voltage sweep will be necessary to turn the current off to a desired level. In this case, the threshold voltage of the device can be reduced, meaning a larger current drive for the same circuit bias voltage, and thereby providing higher performance when the gate and drain are in the active range. The subthreshold slope also determines the residual drain current (I_L) for the same threshold voltage, as shown in the figure. A steeper slope reduces the residual current, if the residual current is mainly due to diffusion of the carriers through the potential barrier between the source and the drain and not due to other factors such as leaky junctions. The subthreshold slope degrades as the transistor channel length is reduced to

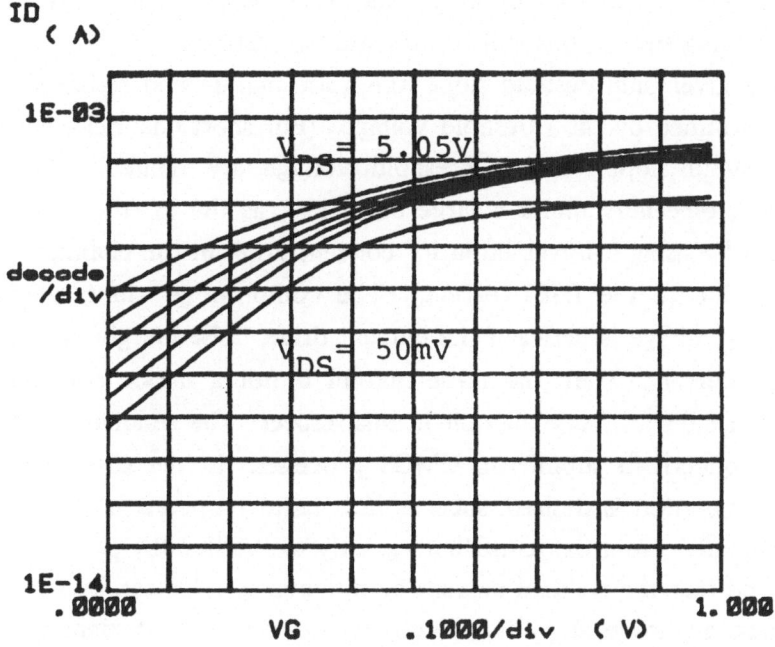

Fig. 6.2 Drain current (log. scale) vs. gate bias for a short channel MOSFET with V_{DS} increasing from 50 mV to 5.05 V.

the extent that short channel effects are significant. Fig. 6.2 shows the subthreshold characteristics of a short channel transistor. The degradation of the slope is more serious for higher drain biases, resulting in significant residual leakage current at zero gate bias. This degradation is due to the drain-induced barrier lowering (DIBL) effect. The physics of the subthreshold characteristics have been discussed in detail in many articles [6.3]-[6.6]. It is useful to point out that in the simplest case where DIBL effects are absent, the subthreshold slope can be approximately expressed as

$$S = 2.3\frac{kT}{q}[1 + \frac{C_D}{C_{ox}}]$$ (6.1)

where C_D and C_{ox} are the capacitance of the depletion layer and the gate oxide capacitance respectively. It can be seen here that by reducing the gate oxide thickness, the subthreshold slope can be reduced significantly. The absolute minimum value of S is 59.6 mV/dec at room temperature. The extraction of the subthreshold for short channel devices using simulations will be presented in section 3.

For a given subthreshold slope, the noise margin and residual current are determined by the threshold voltage. (For short channel devices, the subthreshold slope and threshold voltage are related. The slope degrades, especially under a large drain bias, as the threshold voltage is reduced by using a lower impurity concentration in the channel region. This is due to the DIBL effect.) The compromise in targeting the threshold voltage is between the current drive, noise margin and residual leakage current. Here the noise margin is not a major concern since CMOS circuits are very tolerant in this respect. The residual current is a major concern in submicron CMOS processes. In the case where low power operation is desired, such as in a battery operation environment, the residual current can contribute greatly to the standby power. Thus it is desired that the residual current be minimized. This can be realized by choosing a threshold voltage high enough to prevent significant off current. A higher threshold voltage reduces the residual current not only because of the magnitude of the threshold voltage, which provides more turn-off margin for the transistor, but also because of the higher channel impurity concentrations which reduces the DIBL effects. Hence the subthreshold slope also improves.

In the "turned-on" region, the transistor characteristics can be described by the following simple equations for long channel lengths, typically larger than 3μm, and drain bias below saturation [6.10]:

$$I_{DS} = \frac{W}{L}\mu C_{ox}(V_G - 2\psi_B - \frac{V_D}{2})V_D$$
$$- \frac{2W}{3L}\mu\sqrt{2\varepsilon q N_A}\,[(V_D + 2\psi_B)^{3/2} - (2\psi_B)^{3/2}] \qquad (6.2)$$

$$I_{DS} \simeq \frac{W}{L}\mu C_{ox}[(V_{GS} - V_T)V_{DS} - \frac{1}{2}V_{DS}^2] \qquad (6.3)$$

for low doping concentrations and thin gate oxides;

$$I_{DSAT} = \frac{W}{2L}\mu C_{ox}(V_{GS} - V_T)^2 \qquad (6.4)$$

at saturation. Here μ is the mobility of the carriers in the channel. Obviously it is necessary to have a good mobility model to accurately describe the device characteristics. The mobility is dependent on the vertical electric field in the channel, and for short channel MOSFETs, the effect of the lateral field on the mobility has to be taken into consideration [6.7]. In very short channel transistors, the major deviation from the above equation is the velocity saturation phenomenon, where I_{DS} is not proportional to $(V_{GS}-V_T)^2$ in the saturation region, but rather proportional to $(V_{GS}-V_T)$, as given by the following equation [6.8]:

$$I_{DSAT} = WC_{ox}V_s(V_{GS} - V_T) \qquad (6.5)$$

where V_s is the saturation velocity of the carriers. Fig. 6.3 shows typical transistor I-V characteristics of a long channel MOSFET, and Fig. 6.4 for a short channel device.

It is very useful to be able to predict the performance of the transistors, so that the circuit performance can be predicted before fabrication is actually done. There is always a compromise between transistor performance and process complexity. If the device performance can be predicted, then an optimized process and device design can be developed. There are many other device characteristics issues that the engineer has to be concerned about. One example would be the drain to source breakdown of the MOSFET due to the parasitic bipolar effect between the source and drain [6.9],[6.10]. Fig. 6.5 shows the breakdown behavior of an n-channel transistor for a channel length of one micron. Fig. 6.6 shows the source-drain breakdown voltage (BV_{DS}) versus the effective channel length for n-channel MOSFETs with conventional source/drain structure using arsenic implant. This causes serious reliability problems in the circuit. Many techniques such as using epitaxial silicon on low resistivity substrate, graded junctions, LDD

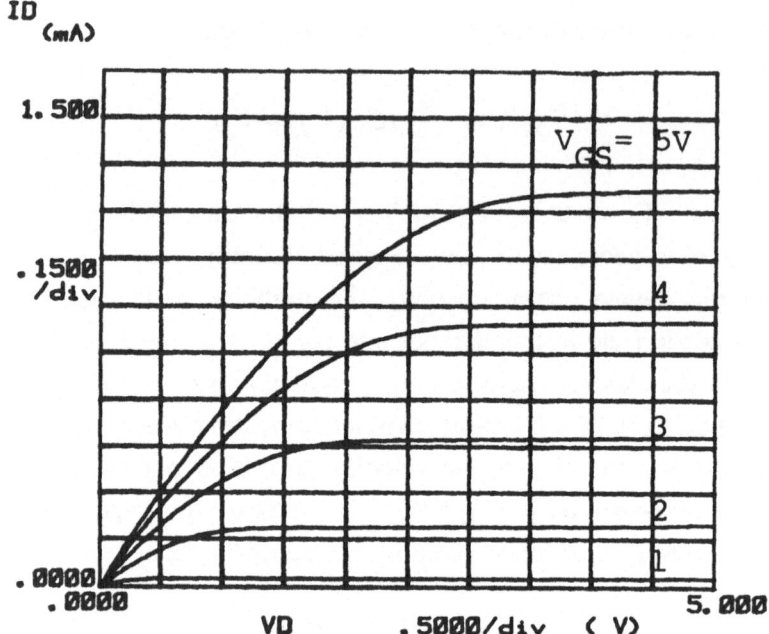

Fig. 6.3 *I-V* characteristics of n-channel MOSFET
with long channel length (~5 μm).

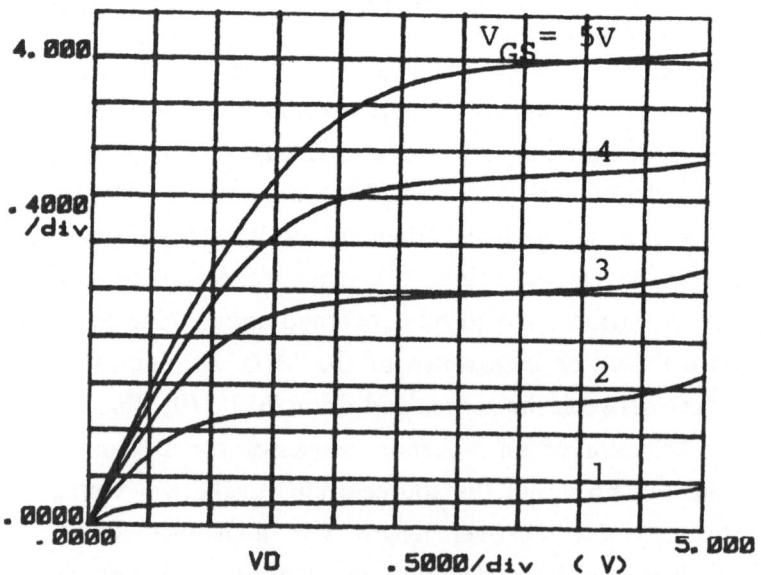

Fig. 6.4 *I-V* characteristics of n-channel MOSFET
with short channel length (~0.7 μm).

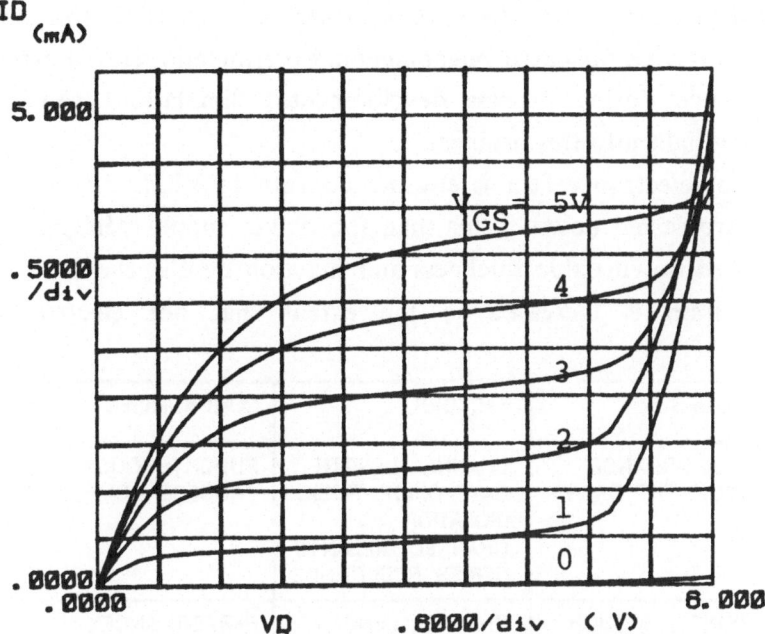

Fig. 6.5 Source-drain breakdown characteristics of n-channel MOSFET with short channel length.

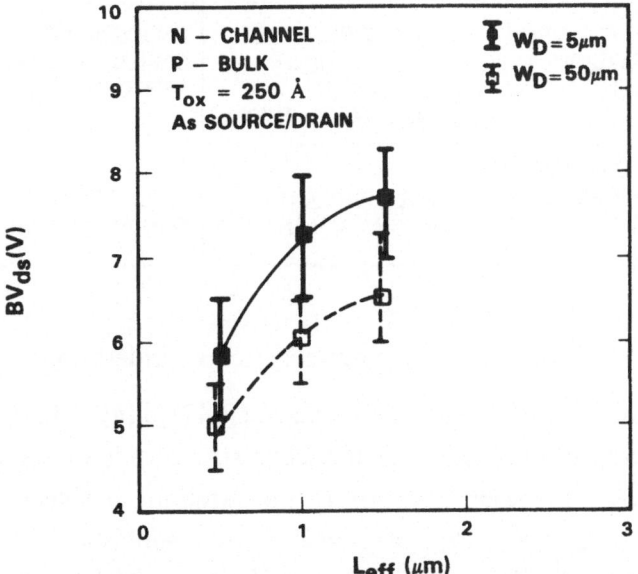

Fig. 6.6 Breakdown voltage (BV_{DS}) of n-channel MOSFET vs. effective channel length (L_{eff}) and width (W_D).

structure [6.11], etc. have been implemented to minimize this problem. Since this is a complicated phenomenon, experimental data is often used as reference during process development. Simulations can provide physical insight into this problem.

The hot electron effect is another concern in VLSI. As the channel lengths are scaled down faster than the power supply voltage, together with scaled down oxide thickness and junction depths, the electric field at the drain is increased to the extent that hot electrons cause

ISSUES	FACTORS	COMPROMISES
PERFORMANCE (CURRENT DRIVE)	CHANNEL LENGTH GATE OXIDE THICK. ISOLATION CHANNEL IMPLANTS SERIES RESISTANCE	PUNCHTHROUGH PROCESS COMPLEXITY
PUNCHTHROUGH	CHANNEL LENGTH GATE OXIDE THICK. GATE MATERIAL CHANNEL IMPLANTS JUNCTION DEPTHS	PERFORMANCE PARASITIC CAPAC. SERIES RESISTANCE BODY EFFECT
HOT ELECTRON/ BREAKDOWN (N–CHANNEL)	S/D STRUCTURE SPACER WIDTH EPI–WAFERS GATE OXIDE THICK.	PROCESS COMPLEXITY SERIES RESISTANCE PERFORMANCE
DIFFUSION CAPAC. PERIPHERAL: AREAL:	FIELD IMPLANT N–CH. IMPLANT	FIELD LEAKAGE PUNCHTHROUGH

Table 6.1 Device optimization considerations

degradation of the device performance [6.12]-[6.14]. In this book, a detailed study of this effect in the LDD structure is presented. Another phenomenon of major concern is the latch-up problem in CMOS. Simplified models [6.15] as well as sophisticated numerical programs [6.16],[6.17] have been used to simulate the latch-up characteristics. This is another example of CAD in device study. Table 6.1 shows the device parameters that have to be optimized (first column), the factors affecting the parameters (second column), and the compromises that have to be made. For example, reducing the channel length will increase the current drive, but may cause punchthrough problems. The table shows that device optimization is a complicated process in VLSI development.

6.2 CAD Tools for Simulation of Device Parameters

In this section, the simulation tools that are used at Hewlett-Packard Laboratory for the simulation of device parameters are discussed. From the previous discussion on the methodology of simulations in process development, it is clear that we need to choose the programs that will provide the necessary information with the desired accuracy in the shortest time. This means that we have to use different programs in different situations, even for generating the same parameter. The most significant differences between devices are long channel versus short channel. The following discussions will provide general guidelines in choosing the programs for device simulations. Note that this section is concerned only with transistor parameters. Chapter 13 will discuss the programs that are appropriate for generating the parasitic resistances and capacitances.

Table 6.2 shows the two categories of programs that are used for process simulations and device simulations. (The SEDAN, SDVICE and PISCES programs have been implemented at Hewlett-Packard Laboratories, but will not be discussed in this book.) In most cases, the simulated impurity profiles and insulator structures can be transferred to the device simulation programs for generating the device parameters. Table 6.3 summarizes the programs that are appropriate for generating the critical parameters for process development. The table is listed in order of increasing complexity, which means increasing computing time. This does not necessarily mean increasing accuracy since it depends on the particular device being considered. The SUPREM program can be used to generate threshold voltages of long channel MOSFETs, or any device that exhibits negligible short channel behavior. This is also the

PROCESS	DEVICE
SUPREM SUPRA SOAP SEDAN	SDVICE GEMINI CADDET SIFCOD PISCES

Table 6.2 Process and device simulation programs used at Hewlett-Packard Laboratories.

PROGRAMS	DEVICE PARAMETERS	REMARKS
SUPREM	VT	LONG CHANNEL, 1—D
GEMINI	VT, SUBVT, VPT	GAUSSIAN PROFILE
SUPREM+GEMINI	VT, SUBVT, VPT	ARB. 1—D PROFILE
SUPRA+GEMINI	VT, SUBVT, VPT	ARB. 2—D PROFILE
SOAP+SUPRA+ GEMINI	VT, SUBVT, VPT	ARB. 2—D PROFILE & ISOLATION
CADDET	VT, SUBVT, VPT, I—V	GAUSSIAN PROFILE
SIFCOD	VT, SUBVT, VPT, I—V, GATE DELAY	GAUSSIAN PROFILE
SUPRA+SIFCOD	VT, SUBVT, VPT, I—V, GATE DELAY	ARB. 2—D PROFILE
SOAP+SUPRA+ SIFCOD	VT, SUBVT, VPT, I—V, GATE DELAY	ARB. 2—D PROFILE & ISOLATION

VT=THRESHOLD VOLTAGE
SUBVT=SUBTHRESHOLD CHARACTERISTICS
I—V=CURRENT—VOLTAGE CHARACTERISTICS ABOVE THRESHOLD
VPT=PUNCHTHROUGH VOLTAGE

Table 6.3 Programs used for generating the device parameters, in order of complexity.

fastest approach to provide a preliminary set of data even for short channels, since it is often possible to make rough corrections on the long channel threshold voltage to provide an estimate of the short channel threshold voltage.

The GEMINI program has the next higher level of complexity. It solves the two-dimensional (2-D) Poisson equation and generates device parameters such as the subthreshold slope. In the case of short channel devices, where the threshold is very sensitive to the channel length and drain bias, it is necessary to use this program to calculate the threshold voltage. In this case, two-dimensional effects are dominant, with the drain bias affecting the potential barrier between the source and drain. Fig. 6.7 shows a simulation of a short channel transistor with a drain bias of 3 V. The depletion region at the drain extends into the channel area. This shows clearly that 1-D analysis is inadequate and 2-D simulation is absolutely essential. The GEMINI program can provide a reasonable approximation of the channel, source and drain profiles, using Gaussian distributions.

The third degree of complexity is the combination of SUPREM and GEMINI, which provides accurate vertical impurity profiles for the channel and the source/drain, during the device simulation. In this case, the channel and source/drain profiles are more realistic, since they are typically not Gaussian. For example, in short channel devices, a combination of shallow and deep implants is used to set the threshold voltage as well as the punchthrough voltage. Also, the source and drain profiles may be much steeper than a Gaussian profile, due to the thermal diffusion properties of the impurities, which depends on the impurity concentrations. [6.18] Fig. 6.8 and Fig. 6.9 show typical channel and source/drain profiles respectively, as simulated by SUPREM. The source/drain impurity profile may not affect the threshold voltage simulation significantly, but will affect the DIBL and hot electron effect simulations.

The fourth degree of complexity is the combination of SUPRA and GEMINI, which provides not only the vertical impurity profile, but also the horizontal distribution at the source/drain. This is essential if one

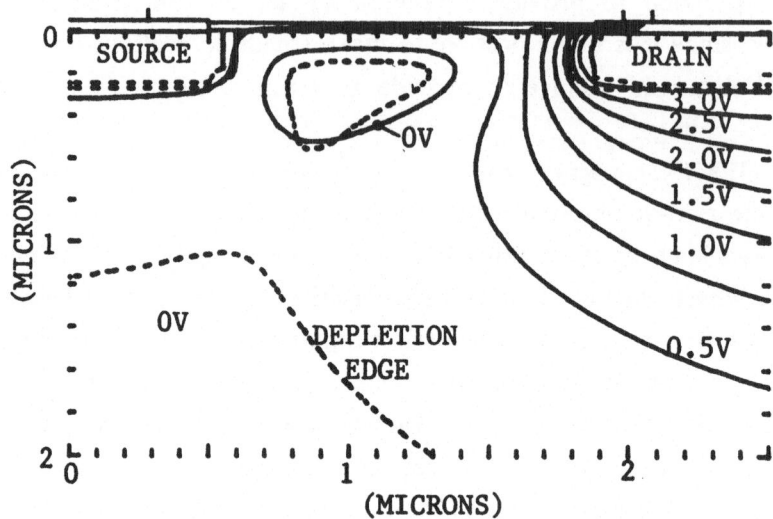

Fig. 6.7 Simulation of an n-channel MOSFET with
channel length of 1.2 μm. The gate and drain
bias are 0 and 3 V respectively.

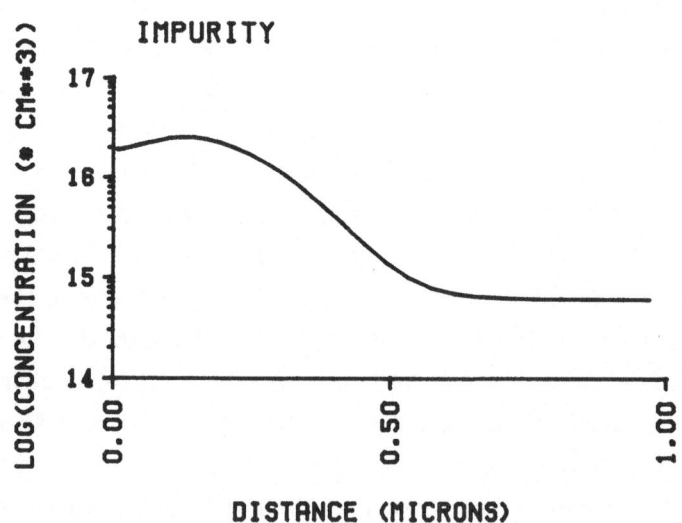

Fig. 6.8 Channel impurity profile of n-channel
MOSFET, with a B11 implant of 5E11 cm^{-2}
at 30 KeV and 4E11 cm^{-2} at 70 KeV.

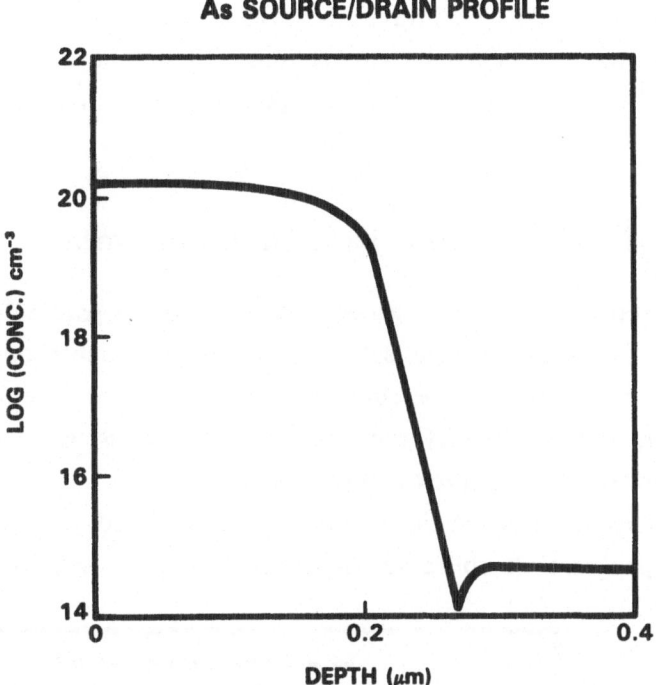

Fig. 6.9 Source/drain profile of n-channel MOSFET

wants to simulate source/drain structures that are more complicated, such as the LDD [6.11] structure. SUPRA also allows one to simulate isolation structures, with certain limitations, and couple that structure to GEMINI for device simulations. This is useful for simulation of narrow width effects or for the simulation of field parasitic transistors with advanced field isolation structures. For more complicated isolation structures such as SWAMI [6.19], it is necessary to couple the SOAP, SUPRA, and GEMINI programs to simulate the narrow width effects.

The CADDET program is used when it is necessary to investigate the full *I-V* characteristics of the device. The program does not accept SUPREM or SUPRA output files. Hence it is not as accurate as the SUPREM-GEMINI or SUPRA-GEMINI combinations in the simulation of the threshold voltage, unless the profiles are carefully specified by first running the SUPREM program and then approximating the profile with a Gaussian distribution. The applications of this program are described in Chapter 3.2, 11 and 12. The SIFCOD program, which is general-shape, 2-D, 2-carrier, transient device simulator, is a relatively new simulation program, hence its applications will not be discussed in this chapter. The reader is referred to Chapter 3.3 for a brief discussion of application examples.

6.3 Methods Of Generating Basic Device Parameters

In this section, methods of generating basic device parameters will be presented. Threshold voltage, subthreshold slope, punchthrough voltage and transconductance will be simulated using different techniques. The assumptions used by the different programs will be discussed, so that the users are aware of the approximations being applied.

The first example is the simulation of the threshold voltage using the SUPREM program. In this case, the threshold voltage will correspond to

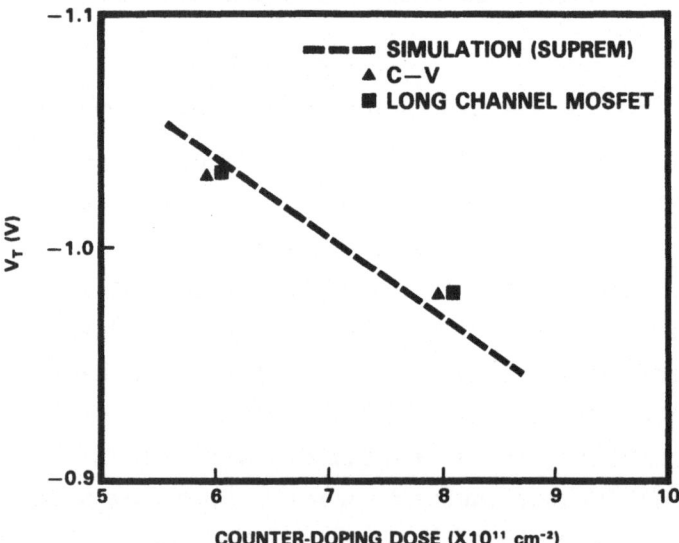

Fig. 6.10 Simulation of p-channel threshold voltage vs.
counter-doping dose.

the MOS capacitor threshold, which is very close to the long channel MOSFET threshold measured by the current versus gate voltage method. The model used is based on the full depletion approximation and the assumption of quasi-neutral impurity profiles [6.20]. The user inputs the process steps for the MOS structure. The program then calculates the impurity profile, and the threshold voltage as a function of the substrate bias. Fig. 6.10 shows the curve of simulated threshold voltages for p-channel transistors as a function of the counter-doping implant dose, together with experimental data. Good agreement is observed. This method of simulating the threshold voltage has the advantage of being the most simple, and the least time consuming. The limitation is that the threshold corresponds to the long channel threshold. This program is useful when one wants to look at the dependence of threshold voltage on the impurity profile without the complication of the short channel effects.

If the threshold voltage of a short channel transistor is to be simulated, then two-dimensional numerical analysis is necessary to provide good accuracy. The combination of SUPREM and GEMINI in most cases provides sufficient accuracy. The channel and source/drain implant profiles are first simulated by the SUPREM program. The structure of the transistor is defined by the GEMINI program. The impurity profiles at the channel and source/drain area are provided by the SUPREM data file after the SUPREM simulation is completed. Fig. 6.11 shows the bird's-eye-view of the impurity profile generated by the GEMINI program using the SUPREM data. The source/drain and channel profiles are displayed very clearly. The GEMINI program then solves Poisson's equation based on those profiles and bias conditions. It is important to know the threshold voltage of a short channel device biased at the power supply voltage V_{DD}, since this is the condition at which the device is biased when it is in the off state in an invertor circuit. For large drain bias, short channel effects are very significant and device characteristics have to be determined by 2-D simulations. The GEMINI program simulates the subthreshold characteristics of the transistor under this bias condition by calculating the drain to source current versus gate bias, from which the subthreshold slope and threshold voltage can be extracted.

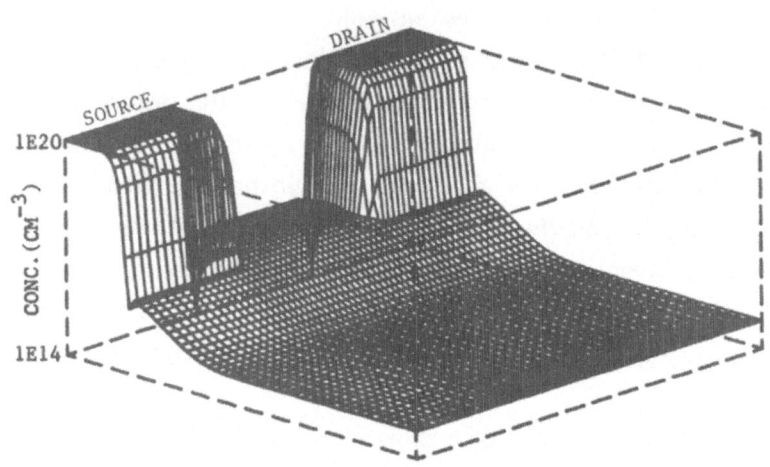

Fig. 6.11 Bird's-eye-view of the impurity profile (log.
scale) for n-channel MOSFET. The substrate
concentration is 6E14 cm^{-3}.

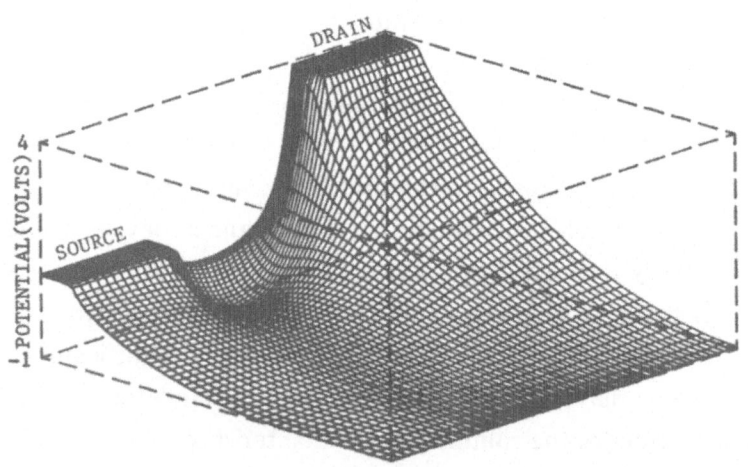

Fig. 6.12 Bird's-eye-view of the potential profile for
n-channel MOSFET, with gate, drain and
substrate bias of 0, 3, and -1 V respectively.

The GEMINI program solves Poisson's equation within the device structure and generates the potential profile. Fig. 6.12 shows the bird's-eye-view of the potential profile within the device structure. The effect of the drain bias in lowering the potential barrier can be observed qualitatively. In the GEMINI program, the quasi-Fermi level at the channel region is set equal to that of the drain bias. Hence under a large drain bias, the program underestimates the inversion charge after the device is turned on, which means that the subthreshold current will continue to increase exponentially even after the threshold voltage is

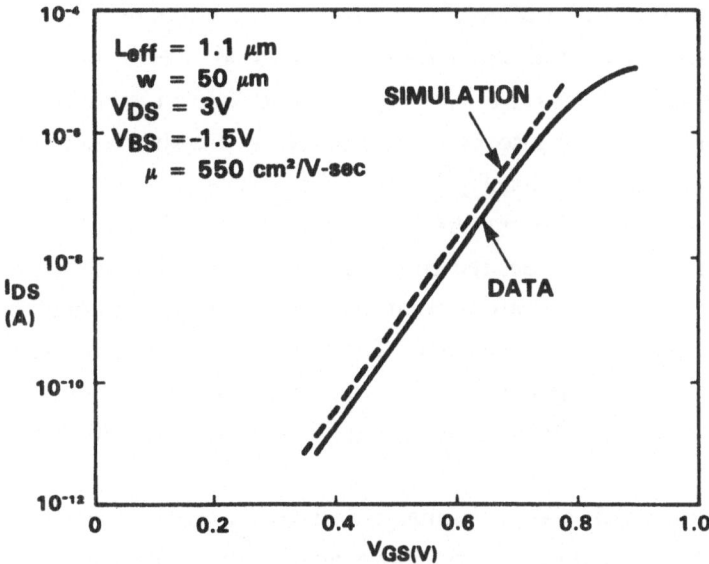

Fig. 6.13 MOSFET subthreshold characteristics simulation.

reached. Fig. 6.13 shows the simulation of the subthreshold characteristics of a n-channel MOSFET, together with the experimental data. The simulated data begins to deviate from the experimental data at high current levels, hence precautions need to be taken when interpreting the simulation results. It is useful to define a threshold current, from which the threshold voltage can be extracted. The conventional value of the threshold current is given by [6.21]

$$I_{TH} = (W_{eff}/L_{eff}) \times 10^{-7} \quad A \tag{6.6}$$

where L_{eff} and W_{eff} are the effective channel length and channel width respectively. If this value of threshold current is used, then the simulated threshold voltage will be too low, due to the deviation from experimental data at turn-on. This deviation is reduced for longer channel lengths. For example, at L_{eff} = 2.5μm, the correction in the simulated threshold voltage is 0.03 V. In order to be consistent with experiment, it would be necessary to include this correction factor for short channel transistors. Another method would be to use a lower value for the threshold current during measurement. Taking the correction into account, the agreement between experiment and simulation is very good over a wide range of implant doses and energies (see Fig. 6.15 and 6.16 in section 4). The subthreshold slope can also be calculated from this simulation by taking the inverse slope of the $\log(I_{DS})$ versus V_{GS} curve.

When the transistor structure is more complicated, such as the LDD structure, or when it is desired to calculate the narrow width effect for transistors with complicated isolation structures, the combination of SUPRA and GEMINI is necessary. The two-dimensional structure of the device will be simulated by SUPRA, and then coupled to the GEMINI program for solving Poisson's equation. This technique will be shown in Chapter 10.

In the design of short channel transistors, a major concern is the problem of "punchthrough", or more accurately, drain-induced barrier lowering (DIBL) [6.4]. The bias applied at the drain of the transistor has the effect of lowering the potential barrier between the source and the drain. "Punchthrough" traditionally means that the drain and source depletion regions are merged, and the maximum potential barrier to carriers is less than the junction "built-in" voltage. In this case, the bias at the drain can affect the potential distribution at the source. But the leakage current may still be negligible until the drain bias has caused significant lowering of the potential barrier height between the source and the channel. Since the diffusion current between the source and the drain is exponentially proportional to this barrier, the leakage current is very sensitive to the drain bias at short channels. The detailed physics of

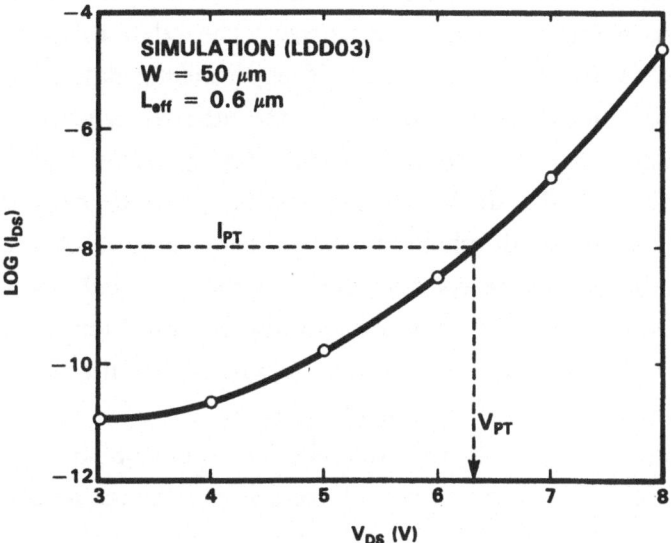

Fig. 6.14 N-channel MOSFET punchthrough characteristics.

DIBL will be discussed in Chapter 7. Since leakage current is a major concern for low power operation, the DIBL effect is always simulated during process development. (This is especially important for VLSI device development where channel length is one micron or less.) This can be accomplished by using the combination of SUPREM-GEMINI or SUPRA-GEMINI. The process simulator generates the channel and source/drain profiles, and the GEMINI program simulates the DIBL effect as a function of the drain bias. Fig. 6.14 shows the simulated data of leakage current at zero gate bias vs. the drain bias, for an n-channel transistor with channel length of 0.6 μm. It is useful to define a "punchthrough voltage", which can be defined as the drain bias at which a certain level of drain current (punchthrough current) is is observed at zero gate bias. The definition of this current level would depend on the application of the devices. Fig. 6.14 shows the determination of the "punchthrough voltage" V_{PT} using 10 nA as a reference current, which is found to be 6.3 V for the device under study.

6.4 Relationship Between Device Characteristics and Process Parameters

In this section, the relationship between device electrical characteristics and process parameters will be discussed. The relationship can best be understood by using a combination of experimental data and simulated data. Simulations allow one to observe the internal structure as well as the internal electrical condition of the device, hence allowing better understanding of the device characteristics. The dependence of the threshold voltage, subthreshold slope, and short channel effects on the channel implant parameters and device structures such as gate oxide thickness and junction depths are discussed. Transistors with channel length down to one micron will be discussed in this chapter, while submicron devices will be discussed in Chapter 8.

To understand the effect of channel implant on the device characteristics such as the threshold voltage, it is better to begin with a longer channel length, such that the correlations between the threshold voltage and implant parameters will not be complicated by significant short channel effects. The simulated threshold voltage as extrapolated from the subthreshold slope for different channel implant profiles for the case of L_{eff} = 2.6 μm and W_{eff} = 48 μm is shown in Fig. (6.15), together with test results from CMOS lots. No corrections in the simulated threshold voltage are necessary since the correction is small at this channel length. The threshold voltages of the data have been normalized to an interface fixed charge density (Q_{SS}) of 2E10 cm^{-2}, to be consistent with that used in the simulations. V_{DS} and V_{BS} are 3 and -1 V respectively. The gate oxide thickness is 35 nm. The horizontal axis indicates the channel implant parameters. The numbers (A,B) mean that the shallow implant (30 KeV) has a dose of AE11cm^{-2}, and the deep implant (50 or 70 KeV) has a dose of BE11cm^{-2}. The profiles are separated into two groups, one with a deep implant energy of 50 KeV, and the other of 70 KeV. The figure shows that the simulation and experiment agree well for a large range of implant parameters. It also must be emphasized that in most cases, the comparison of experimental

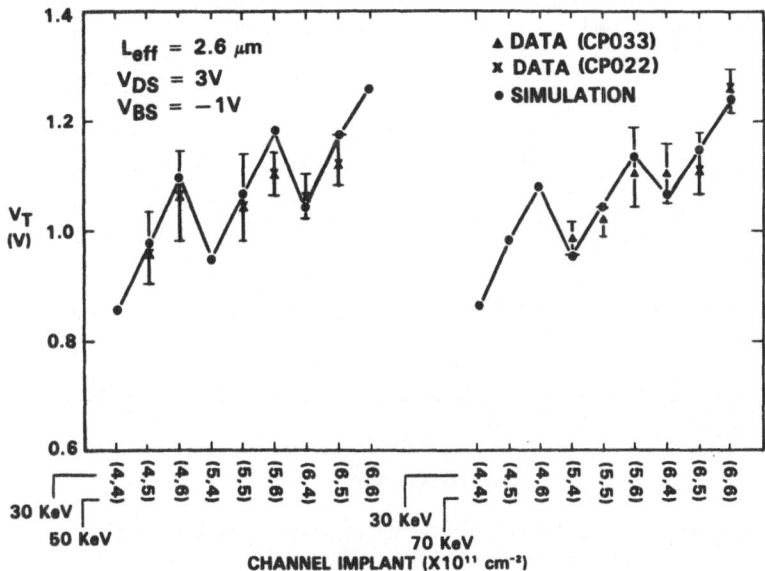

Fig. 6.15 Threshold voltage vs. channel implant for
n-channel MOSFET with channel length of
2.6 μm.

data and simulation data is meaningful only when there are significant statistics, due to process variations and measurement precision. From the figure, it can be seen that for the same implant energy, the threshold voltage increases with dose, as is expected. Also, the dependence of V_T on the shallow implant dose and the deep implant dose are approximately equal. The increase in V_T is approximately 0.1 V per 1E11cm^{-2} increase in dose, with an accuracy of about 0.02 V. This is because the two doses are similar in magnitude (in contrast to thin oxide devices where the low energy dose is higher than the high energy dose; see Chapter 8), and also because the diffusion of the impurities during the fabrication process has smoothed out the distribution. Fig. 6.8 shows that for a channel boron implant of 5E11 cm^{-2} at 30 KeV and 4E11 cm^{-2} at 70 KeV, and with typical CMOS process temperature cycles, the impurity profile is almost Gaussian, instead of a two peaked structure.

The dependence of V_T on channel profile is also studied for the case of short channel devices. Fig. 6.16 shows the simulated threshold voltages, together with experimental data, for the same range of implant

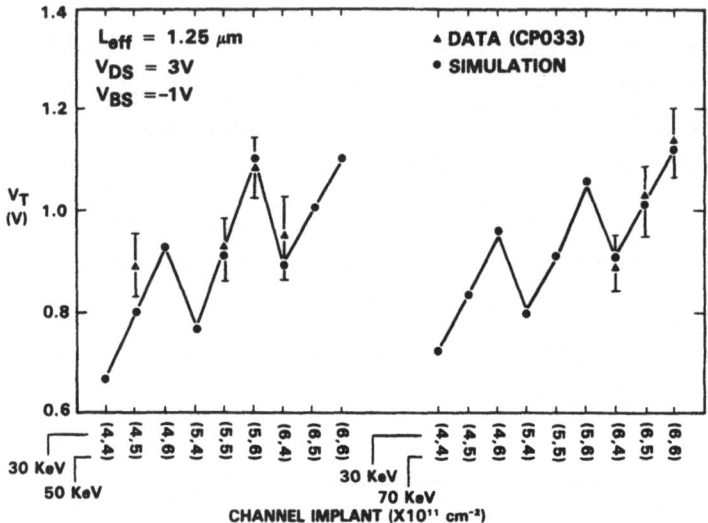

Fig. 6.16 Threshold voltage vs. channel implant profile
for n-channel MOSFET with L_{eff} = 1.25 μm.

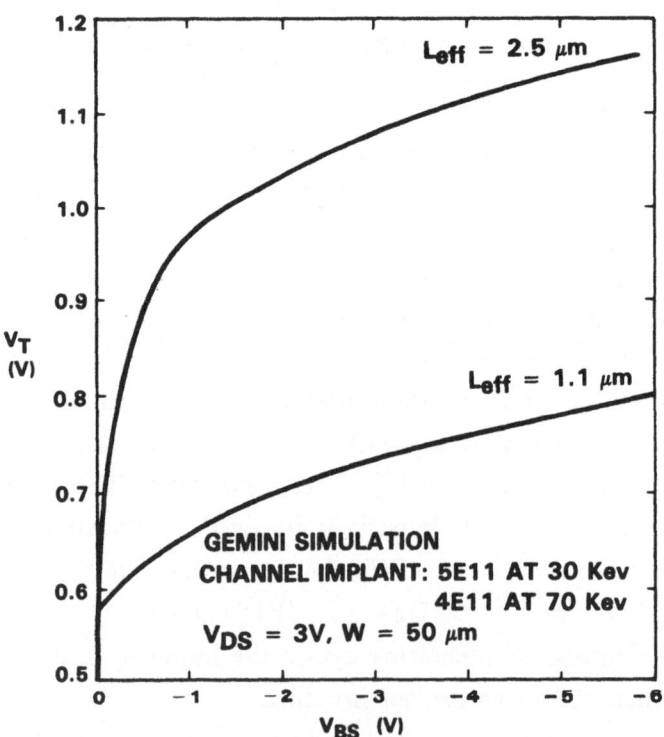

Fig. 6.17 Simulation of the body effect for long and
short n-channel MOSFETs.

parameter values. The simulated threshold voltage is generated from the subthreshold slope simulation, using the SUPREM-GEMINI programs, then corrected for the deviation due to the GEMINI approximation. The agreement between experiment and simulation is quite good. The simulations show that the dependence of the threshold voltage on the implant dose is similar to the device with L_{eff} = 2.5µm.

The sensitivity of the threshold voltage to the substrate bias (the "body effect") can also be studied using simulations. Substrate bias can be applied to the device in the GEMINI program. The threshold voltage, under a desired drain bias, can then be calculated as a function of the substrate bias. Fig. 6.17 shows such a simulation, where the drain bias is at 3 V, for channel lengths of 2.5 µm and 1.1 µm. The results show that the threshold voltage rises rapidly during the first one volt of substrate bias and then rises at a slower rate with increasing substrate bias. This can be explained by the channel implant profile. As the substrate bias is increased, the depletion edge moves from the high impurity concentration region near the surface to the lower concentration region deeper into the substrate. Since the body effect is proportional to the square root of the impurity concentration, the body effect is reduced at higher substrate bias. Another interesting result is that the device with short channel length has significantly less body effect. This can be explained by Fig. 6.18a and 6.18b where the 2-D potential contours and the depletion regions of the two devices are shown at a substrate bias of -1 V and drain bias of 3 V. The depletion region at the drain of the short channel device encroaches into the channel area. This causes an effective reduction of the channel impurity concentration, hence a lower body effect. Note that the channel depletion edge of the short channel device is deeper than that of the long channel one. Fig. 6.19 shows the experimental data for similar devices. The agreement between experiment and simulation is good.

The simulation of the threshold voltage versus L_{eff} is also useful because it defines the appropriate channel length for a particular process. It also shows the sensitivity of the process and device design to short channel effects. The channel length should be chosen such that the

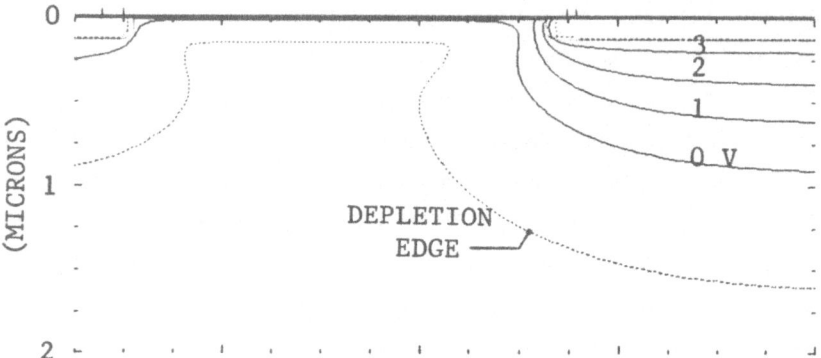

Fig. 6.18a Potential profile and depletion edge simulation of long channel MOSFET, with drain and substrate bias of 3 and -1 V respectively. L_{eff} = 4 μm.

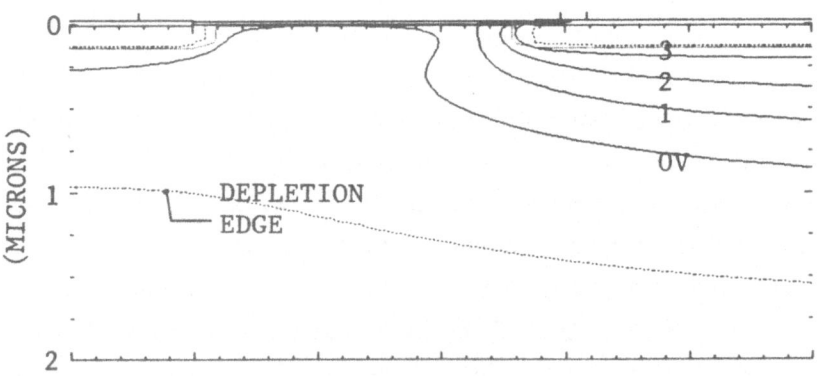

Fig. 6.18b Potential profile and depletion edge simulation of short channel MOSFET with channel length ~ 1.2 μm, with same bias conditions as above.

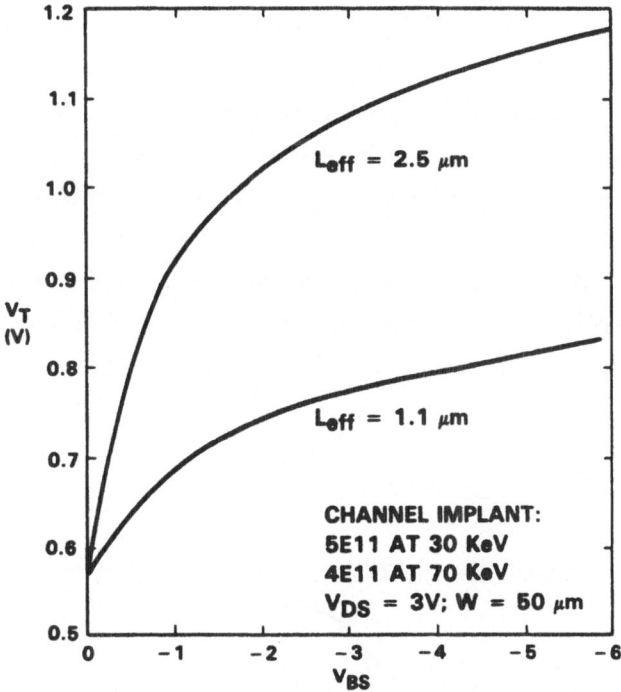

Fig. 6.19 Measurement of body effect for n-channel MOSFETs.

sensitivity of the threshold voltage to the channel length is small, so that any variations in the channel length due to processing will not cause a serious variation in the threshold voltage. The physics of threshold voltage drop with reducing channel length have been discussed in many papers and will not be reproduced here [6.4],[6.22]-[6.23]. The most simple explanation is the charge sharing model [6.22]. In this model, the depletion layer at the drain and source have consumed a significant fraction of the channel charge, thus effectively reducing the channel doping concentration. This is especially severe as the channel length decreases, and when the drain is under a large bias.

Fig. 6.20 shows the simulation of the threshold voltage of n-channel transistors as a function of effective (electrical) channel length, for a drain bias of 3V. The simulations show that for this particular process, the threshold voltage is not very sensitive to the channel length at 1 μm, but is quite sensitive at 0.5 μm.

The above discussions have emphasized the threshold and subthreshold regions of the transistor characteristics. In these regions, short channel effects such as subthreshold leakage and threshold sensitivity to channel length are very significant. The threshold voltage is a very important parameter because it influences the performance (current drive) and also the subthreshold leakage of the device. Therefore, simulations in these regions are always necessary in VLSI device development to assure that the device will perform properly at the desired channel length. The full *I-V* characteristics of the transistor can be simulated by using the CADDET and SIFCOD programs. The reader is referred to Chapter 3 and Chapter 12.

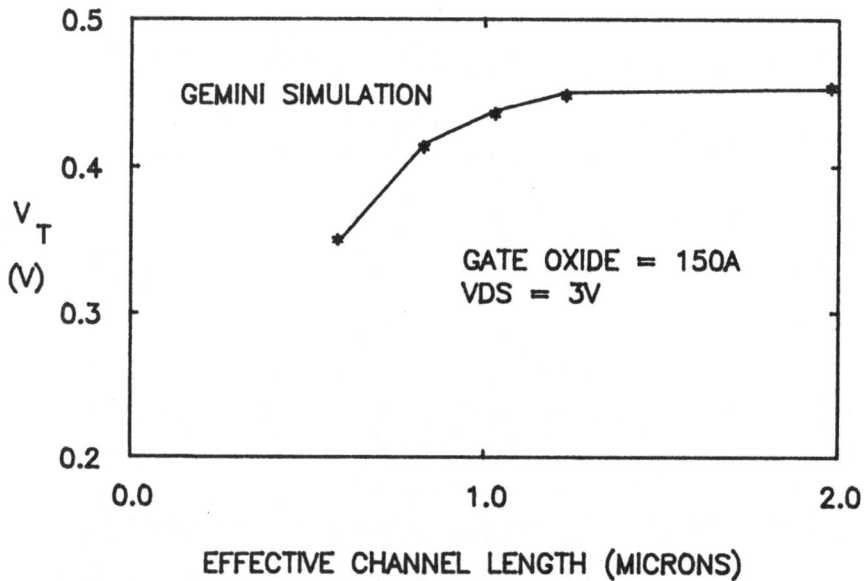

EFFECTIVE CHANNEL LENGTH (MICRONS)

Fig. 6.20 Simulation of threshold voltage vs. effective
channel length for n-channel MOSFET.

References:

[6.1] S. M. Sze, *Physics of Semiconductor Devices*, 2nd ed., New York: Wiley-Interscience, 1981.

[6.2] A. S. Grove, *Physics and Technology of Semiconductor Devices*, New

York: John Wiley & Sons, Inc., 1967.

[6.3] R. R. Troutman, "Subthreshold Design Considerations for Insulated Gate Field Effect Transistors," *IEEE J. Solid State Circuits*," <u>SC-9</u>, April 1974, pp. 55-60.

[6.4] R. R. Troutman, "VLSI Limitations from Drain-Induced Barrier Lowering," *IEEE Trans. Electron Devices*, <u>ED-27</u>, April 1979, pp. 461-468.

[6.5] B. Eitan and D. Frohman-Bentchkowsky, "Surface Conduction in Short-channel MOS Devices as a Limition to VLSI Scaling," *IEEE Trans. Electron Devices*, <u>ED-29</u>, Feb. 1982, pp. 254-266.

[6.6] G. W. Taylor, "Subthreshold Conduction in MOSFETs," *IEEE Trans. Electron Devices*, <u>ED-25</u>, March 1978, pp. 337-350.

[6.7] K. Yamaguchi, "A Mobility Model for Carriers in the MOS Inversion Layers," *IEEE Trans. on Electron Devices*, <u>ED-30</u>, pp. 658-663, June 1983.

[6.8] Y. El-Mansy, "MOS Device and Technology Constraints in VLSI," *IEEE Trans. on Electron Devices*, <u>ED-29</u>, Apr 1982, pp. 567-573.

[6.9] S. E. Laux and F. H. Gaensslen, "A Study of Avalanche Breakdown in Scaled n-MOSFETs," *Tech. Digest of IEDM 1984*, pp. 84-86.

[6.10] F.-C. Hsu, P.-K. Ko, S. Tam, C. Hu and R. S. Muller, "An Analytical Breakdown Model for Short-Channel MOSFETs," *IEEE Trans. Electron Devices*, <u>ED-29</u>, Nov. 1982, pp. 1735-1740.

[6.11] H. Katto, K. Okuyama, S. Meguro, R. Nagai and S. Ikeda, "Hot Carrier Degradation Modes and Optimization of LDD MOSFETs," *Tech Digest of IEDM 1984*, pp. 774-777.

[6.12] L. A. Akers, M. A. Holly and C. Lund, "Hot Carriers in Small Geometry CMOS," *Tech. Digest of IEDM 1984*, pp. 80-83.

[6.13] C. Hu, "Hot Electron Effects in MOSFETs," *Tech. Digest of IEDM 1983*, pp. 176-181.

[6.14] F.-C. Hsu and K.-Y. Chiu, "A Comparative Study of Tunnelling, Substrate Hot-Electron and Channel Hot Electron Injection Induced Degradation in Thin Gate MOSFETs," *Tech. Digest of IEDM 1984*, pp. 96-99.

[6.15] D. B. Estreich, "The Physics and Modeling of Latch-Up in CMOS Integrated Circuits," Stanford Electronics Labs Report #G-201-9, 1980.

[6.16] R. R. Troutman, "Recent Developments in CMOS Latchup," *Tech. Digest of IEDM 1984*, pp. 296-299.

[6.17] M. R. Pinto, R. W. Dutton, H. Iwai and C. S. Rafferty, "Computer-Aids for Analysis and Scaling of Electron Devices," *Tech. Digest of IEDM 1984*, pp. 288-291.

[6.18] D. A. Antoniadis, S. E. Hansen, and R. W. Dutton, "SUPREM II - A Program for IC Process Modeling and Simulation," TR 5019.2, Stanford Electronics Laboratories, Stanford University, Calif., June 1979.

[6.19] K. Y. Chiu, J. L. Moll, K. M. Cham, J. Lin, C. Lage, S. Angelos, and R. Tillman, "The Sloped-Wall SWAMI -- A Defect-Free Zero Bird's Beak Local Oxidation Process For Scaled VLSI Technology," *IEEE Trans. Electron Devices*, ED-30, Nov. 1983, pp. 1506-1510.

[6.20] R. R. Troutman, "Ion-Implanted Threshold Tailoring for Insulated Gate Field-Effect Transistors," *IEEE Trans. Electron Devices*, ED-24, Mar 1977, pp. 182-192.

[6.21] H.-G. Lee, S.-Y. Oh and G. Fuller, "A Simple and Accurate Method to Measure the Threshold Voltage of an Enhancement-Mode MOSFET," *IEEE Trans. Electron Devices*, ED-29, Feb. 1982, pp. 346-348.

[6.22] P. K. Chatterjee and J. E. Leiss, "An Analytic Charge-Sharing Predictor Model for Submicron MOSFETs," *Tech. Digest of IEDM 1980*, pp. 28-33.

[6.23] K. Yokoyama et al., "Threshold-Sensitivity Minimization of Short-Channel MOSFET's by Computer Simulation," *IEEE Trans. Electron Devices*, ED-27, Aug 1980, pp. 1509-1514.

Chapter 7

Drain-Induced Barrier Lowering In Short Channel Transistors

Drain-induced barrier lowering (DIBL)[7.1]-[7.6] has been studied by many workers. The result of DIBL is an increase in the residual leakage current in short channel devices as the drain to source voltage is

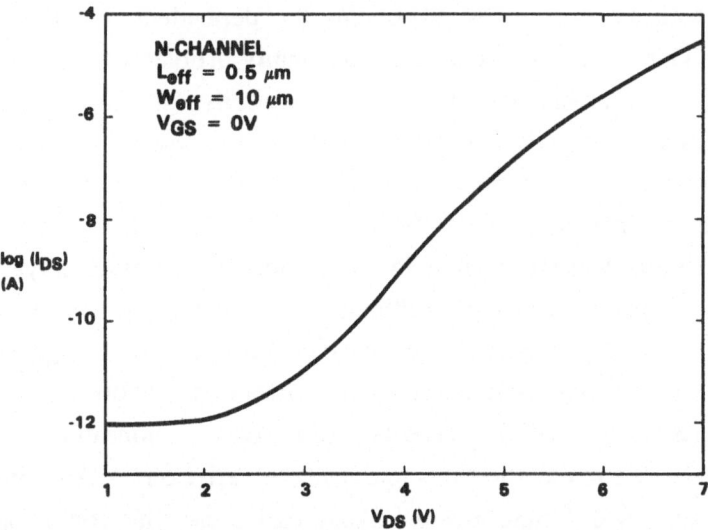

Fig. 7.1 Punchthrough behavior of short channel MOSFET.

increased. Fig. 7.1 shows the measurement of the drain to source current of a short channel MOSFET, as a function of the drain bias, for gate bias of 0V. Note that the current increases exponentially with drain bias. Fig. 7.2 shows the simulated potential profile between the source and

159

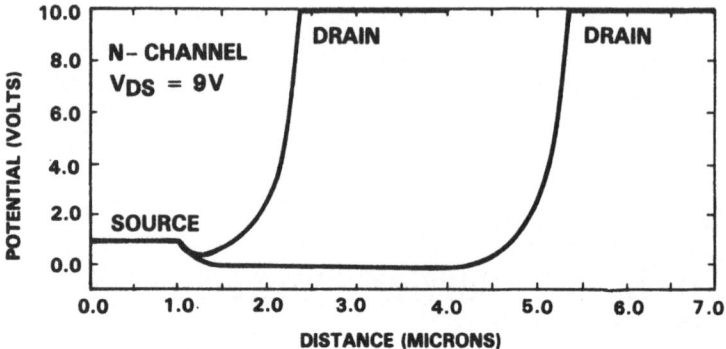

Fig. 7.2 DIBL of long and short channel MOSFETs.

drain of a long and short channel MOSFET, under a drain to source bias
of 9 V. The potential barrier between the source and the channel is
lowered by the drain bias, for the short channel device. The drain to
source leakage current is exponentially dependent on the potential
barrier. This leakage current can cause many problems in circuits such as
dynamic memories or low power circuits in battery operation
environments.   In the first example, if the pass transistor of the
one-transistor dynamic memory cell [7.7] has significant leakage current,
then the bit information of the cell may be lost. In the case of low
power circuits, leakage current in the devices means much larger standby
power.   If severe leakage problems are present, the circuit may not
function properly, especially for NMOS circuits. The design of analog
circuits such as sense amplifiers in memories depends on uniform device
characteristics.   DIBL   effects   can   cause   non-uniform   device
characteristics such as the threshold voltage variations.   Also, in the case
of dynamic logic design, severe leakage can cause the voltage at a node
to drop, causing circuit malfunctions.

Since DIBL is of a major concern in the design of short channel
transistors, simulations are always performed for the process and device
design to make sure that DIBL is not significant, or that the amount of
threshold voltage variation or residual leakage current is tolerable for
the application. The simulations are performed by the combination of
SUPREM, which simulates the impurity profiles, and GEMINI, which

simulates the DIBL effect for the device structure under study. GEMINI also allows the simulation of LDD structure. If the device has a very special source/drain structure, the combination of SUPRA and GEMINI is recommended, but with the restriction that all dopants introduced through the drain "window" are of the same type. In this case, SUPRA generates the source/drain structure, which is then coupled to the GEMINI program for solving Poisson's equation. The following simulations are for transistors with conventional source/drain structures.

The term "punchthrough voltage" (V_{PT}) is used to describe the drain voltage at which the channel current is equal to a reference value, typically in the nA region, when the gate is biased in the off state, usually at zero volt. Before looking into the dependence of the punchthrough voltage on various process and device parameters, it is illustrative to study the effect of the drain bias on the potential barrier between the source and drain using simulations. Fig. 7.3 shows a series of two-dimensional potential contour plots for increasing values of drain bias, for an effective channel length of 1μm. The effect of the drain bias is to reduce the potential barrier between the source and the substrate. The spread of the potential contours towards the source is clear in these figures as the drain bias is increased. For this device, the drain to source current is 1 nA for a width of 50 μm when the drain bias is 9 V. Since the GEMINI program solves only the Poisson equation and does not solve the continuity equation, the calculations are accurate only in a limited range of the current-voltage characteristics. Hence, DIBL simulations should be performed only for low level currents when GEMINI is used. The lowering of the potential barrier height can be seen clearly in Fig. 7.4 where the potential along the current path is plotted for the region near the source. From here on, the term "potential barrier height" refers to the absolute value of the difference between the source potential and the minimum electron potential (or maximum hole potential) along the current path being considered. The current path can be at the surface or in the bulk under these bias conditions. The path that contains the lower potential barrier height will control the punchthrough current. Simulations and experiments have shown that

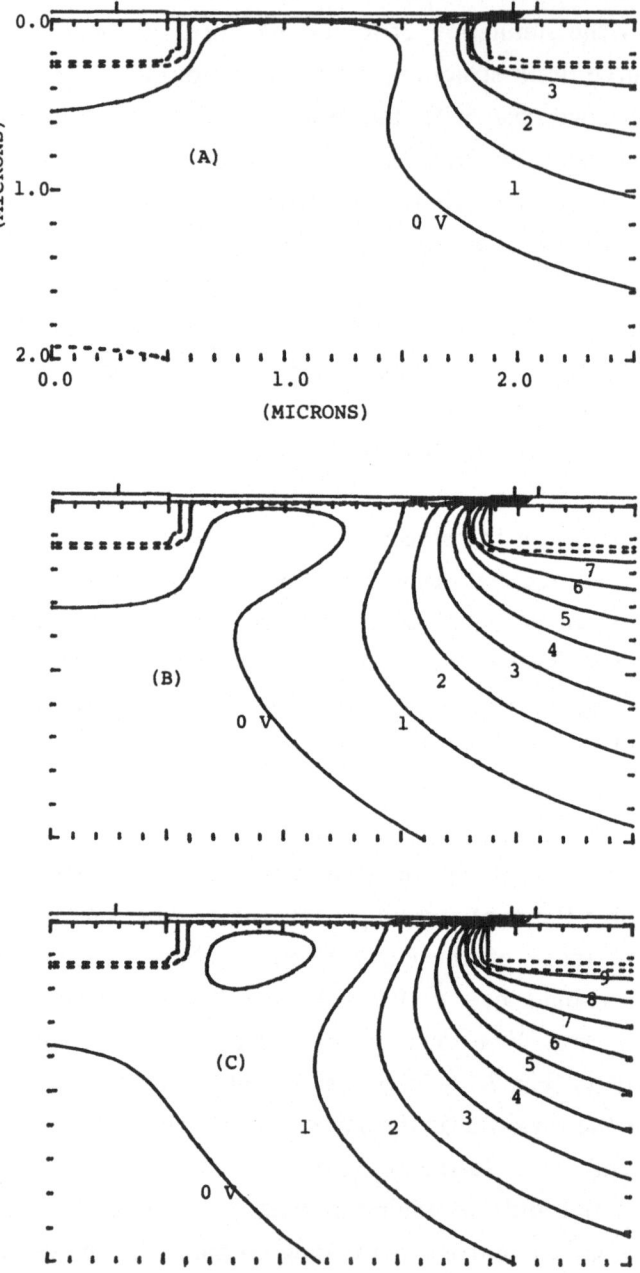

Fig. 7.3 2-D potential profile of n-channel MOSFET
       with drain bias of (A) 3 V, (B) 7 V, (C) 9 V.
       Channel length = 1 μm.

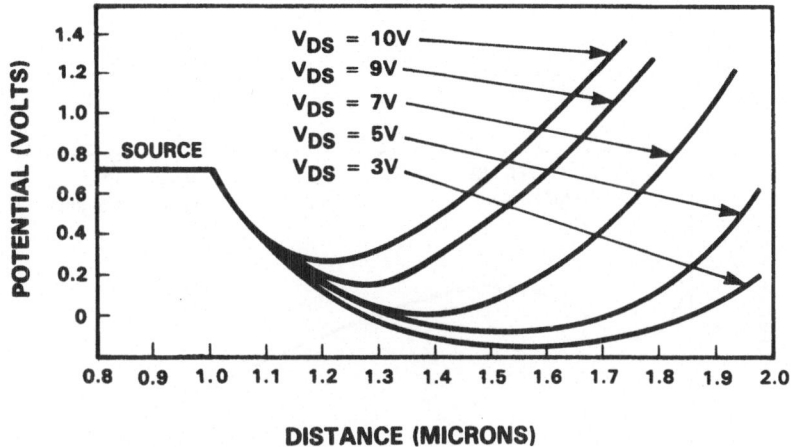

Fig. 7.4 DIBL vs. drain bias for short channel MOSFET.

punchthrough occurs at the surface for lower drain bias, and in the bulk at high drain bias. The following discussion will study this phenomenon in detail.

2-D simulations have been used extensively to understand the punchthrough problem for short channel MOSFETs. Fig. 7.5 shows the potential as a function of vertical distance from the silicon surface into the bulk, at a distance of 0.2 μm away from the source, for the case of drain bias of 9 V. This distance is chosen because it is close to the point where punchthrough occurs. There are two local potential maxima (hence energy minima for electrons), at the surface and in the bulk. Using this figure, one can locate the possible surface and bulk current paths. These possible current paths are plotted in Fig. 7.6, together with the potential contours. The punchthrough current will flow along the path which contains the minimum potential barrier height. Conventionally, the term "saddle point" is referred to the point along the current path where the electron (hole) potential is a minimum (maximum). (The "saddle" shape can be visualized by combining figures 7.4, 7.5 and 7.6.)

In this particular case, according to the simulation results, the punchthrough path occurs at the surface, and the minimum electron potential occurs at $x = 1.24\mu m$ (point A in Fig. 7.6) for a punchthrough current of 1 nA and device width of 50 μm. From here on, the term

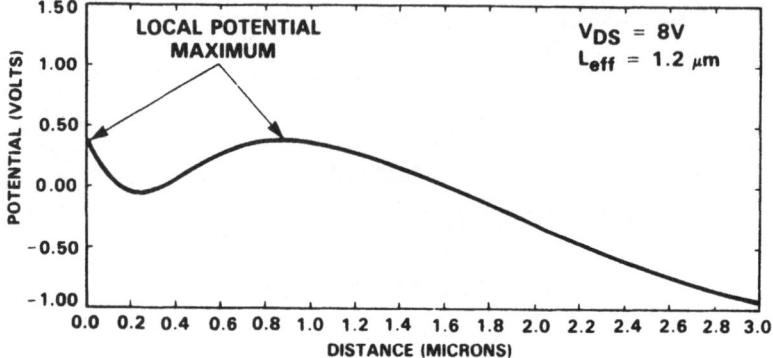

Fig. 7.5 Potential profile of n-channel MOSFET vs.
vertical distance from silicon surface into the
substrate.

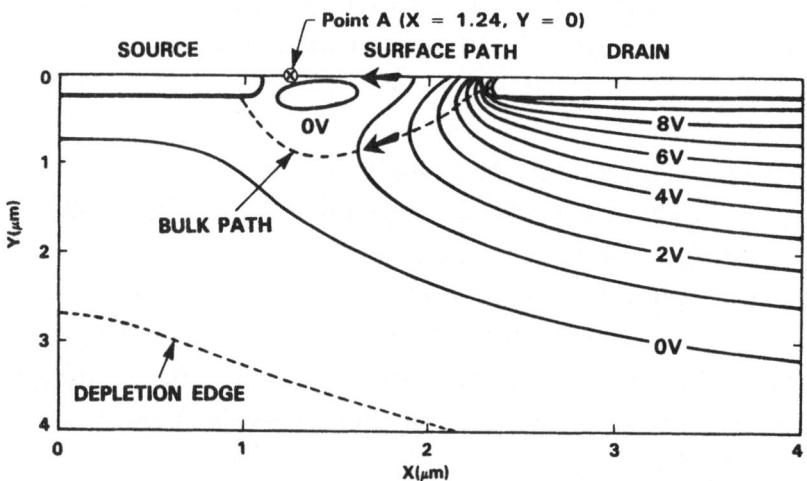

Fig. 7.6 2-D simulation of the potential profile of an
n-channel MOSFET, with a gate and drain
bias of 0 and 9 V respectively. The surface
and bulk punchthrough paths are indicated.

(A)

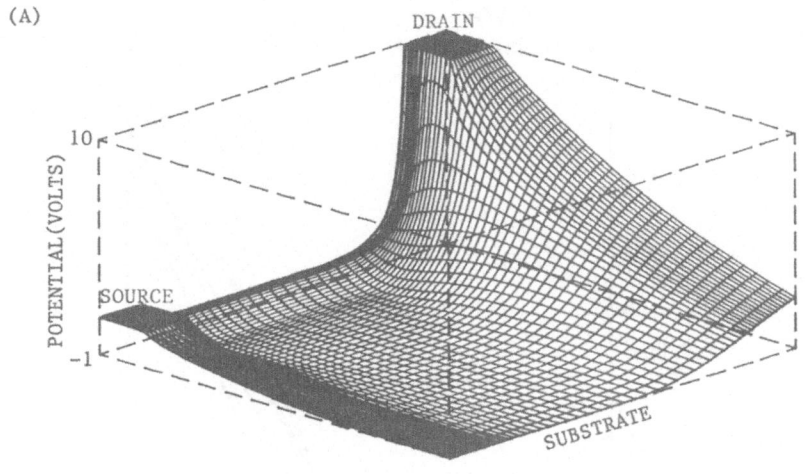

(B)

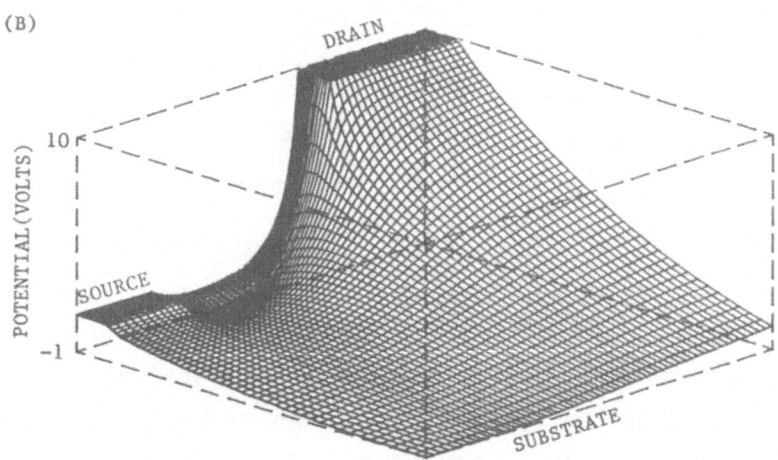

Fig. 7.7 Bird's-eye-view of the potential profile for a long channel (A) and short channel (B) MOSFET, with a drain and gate bias of 9 and 0 V respectively.

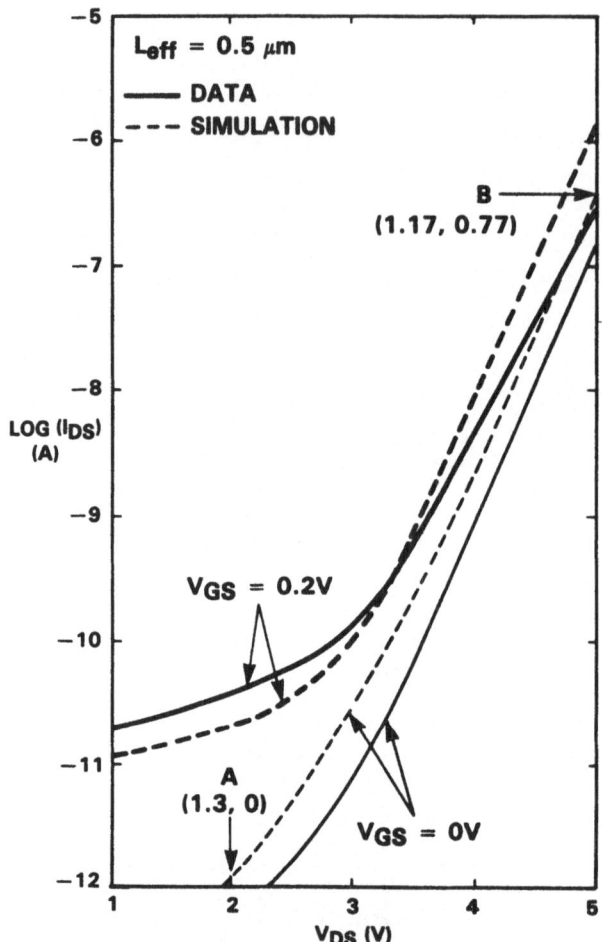

Fig. 7.8 Punchthrough characteristics of short channel
MOSFET, simulation and experimental data.

"punchthrough point" refers to the location of the minimum electron
potential along the punchthrough path.   Fig. 7.7 compares the
bird's-eye-view of the potential profile for   long and short channel
MOSFETs. The effect of the drain bias on the potential profile at the
channel region is evident for the short channel device.

The I_{DS} vs. V_{DS} curves can provide physical insights into the DIBL
behavior of short channel devices. Fig. 7.8 shows the test and simulated
data of n-channel devices, with $L_{eff} = 0.5$ µm.   Log(I_{DS}) is plotted

versus V_{DS} for two values of gate to source voltage (V_{GS}), 0 V and 0.2 V. At high drain bias, the drain current increases rapidly and exponentially with the drain voltage in both cases. Also, the drain currents are approximately equal for the two different gate biases. At lower drain bias, the drain current is less dependent on the drain bias, especially for the case with gate bias of 0.2 V. Also, the dependence of the drain current on the gate bias increases. This behavior is due to a change in the punchthrough current path as drain bias increases. The slope is less steep at voltages less than 3 V. This is because the current path is along the surface. The dependence of I_{DS} on V_{DS} is expected to be less since the surface potential is mainly controlled by the gate bias. Surface punchthrough occurs before bulk punchthrough for the case of $V_{GS} = 0$ V, due to the surface energy band bending, the combination of channel implants, and shallow junctions, producing a lower potential barrier at the surface. The change from surface punchthrough to bulk punchthrough can be more clearly seen for the case of $V_{GS} = 0.2$ V. The dependence of the drain current on the gate bias is also expected to be large for surface punchthrough, as is confirmed by the experimental and simulated data. As the drain bias is increased, the barrier in the bulk is lowered, and eventually changes the current path. Since the drain bias has the major effect on the potential in the substrate, the dependence of I_{DS} on V_{DS} is much stronger in this case, as shown by the data with the positive gate bias.

The simulations allow the punchthrough path to be identified. The punchthrough points (x,y) are indicated for two drain biases for the simulated curve with $V_{GS} = 0$ V. The parameter x is the horizontal distance from the left boundary of the device structure defined in the GEMINI input file. Here the source is located between 0 and 1 µm. The parameter y indicates the distance of the punchthrough point measured from the silicon surface. This point changes from being at the surface to 0.77 µm into the bulk for the drain bias increasing from 2 V to 5 V. This agrees with the previous discussions which concludes that surface punchthrough is occurring at point A and bulk punchthrough is occurring at point B. The potential profile as a function of the vertical

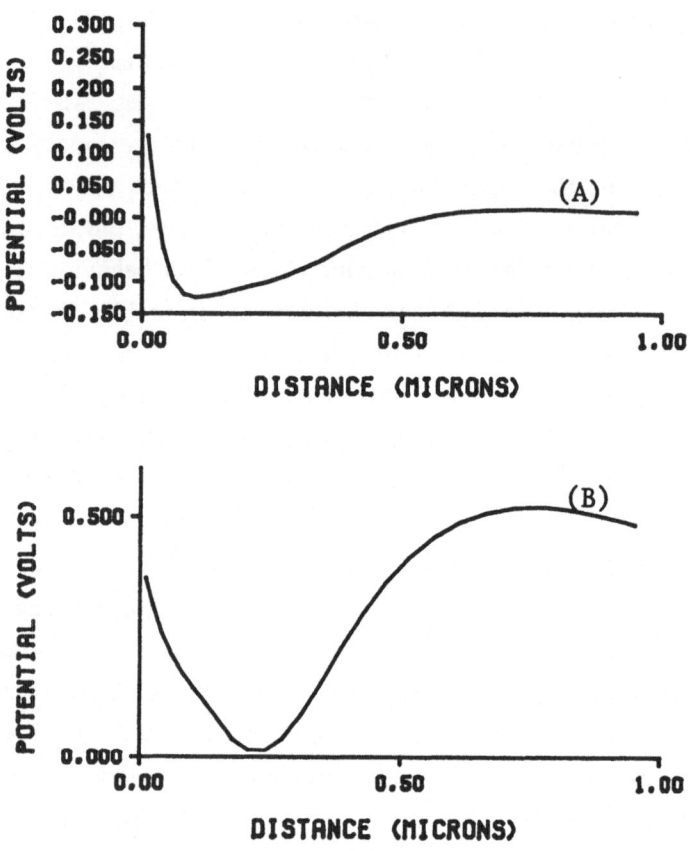

Fig. 7.9 Potential profile vs. vertical distance for
points (A) and (B) in Fig. 7.8.

distance from the surface into the substrate is plotted in Fig. 7.9, for the
two bias points A and B. The horizontal distance from the source is
chosen to be equal to the simulated horizontal position of the
punchthrough point. The potential maximum (hence energy minimum for
electrons) can be seen to be at the surface for bias point A and at the
bulk for bias point B. This identifies the surface and bulk punchthrough
for the two cases. Overall, the simulations agree with the experiment
very well qualitatively. To agree quantitatively well with experimental
data, one needs very precise process simulations of the channel and
source/drain profile, and also very accurate experimental data such as the
effective channel length. Therefore, very good quantitative agreements

for short channel devices at high drain biases are usually not obtained. But still one can gain a lot of insight and have good estimates on the punchthrough behavior of submicron channel length MOSFETs.

References

[7.1] R. R. Troutman, "VLSI Limitations from Drain-Induced Barrier Lowering," *IEEE Trans. Electron Devices*, ED-27, April 1979, pp. 461-468.

[7.2] R. R. Troutman, "Subthreshold Design Considerations for Insulated Field Effect Transistors," *IEEE J. Solid State Circuits*, SC-9, April 1974, pp. 55-60.

[7.3] B. Eitan and D. Frohman-Bentchkowsky, "Surface Conduction in Short Channel MOS Devices as a Limitation to VLSI Scaling," *IEEE Trans. Electron Devices*, ED-26, April 1979, pp. 254-266

[7.4] H. Masuda, M. Nakai, and M. Kubo, "Characteristics and Limitation of Scaled-Down MOSFET's Due to Two-Dimensional Field Effect," *IEEE Trans. Electron Devices*, ED-26, June 1979, pp. 980-986.

[7.5] K. M. Cham and S. Y. Chiang, "Device Design for the Submicrometer P-Channel FET with n+ Polysilicon Gate," *IEEE Trans. Electron Devices*, ED-31, July 1984, pp. 964-968.

[7.6] J. J. Barnes, K. Shimohigashi, and R. Dutton, "Short-Channel MOSFET's in the Punchthrough Current Mode," *IEEE Trans. Electron Devices*, ED-26, April 1979, pp. 446-453

[7.7] Y. A. El-Mansy and R. A. Burghard, "Design Parameters of the Hi-C DRAM Cell," *IEEE J. Solid-State Circuits*, SC-17, Oct 1982, pp. 951-956.

Chapter 8

Transistor Design for Submicron CMOS Technology

In this chapter, the design of transistors for submicron CMOS technology will be presented. The advantages of, as well as issues involved in CMOS technology will first be discussed. Then the concerns for the design of n and p-channel MOSFETs with submicron channel lengths will be discussed. Using simulations, the values of the critical device parameters are determined which will minimize leakage problems in submicron transistors.

8.1 Introduction to Submicron CMOS Technology

Before getting into the details of process development of CMOS circuits using CAD tools, it is useful to understand why we should put so much effort into CMOS. CMOS technology has gained more and more significance, overtaking NMOS technology as the era of VLSI approaches. There are several major reasons. First, when the number of transistors on a chip is approaching a million, the power consumption becomes a limiting factor for VLSI. For NMOS technology, creative circuit design is necessary to minimize power consumption, and expensive packaging is needed for heat dissipation. CMOS circuits, due to the complementary nature of the n- and p-channel transistors, consume much less power, especially in the standby state, where the power consumption is determined by the leakage current of the transistors biased in the off state. The power consumption of the circuit in the active state depends on the switching frequency of the circuits. For example, in the case of 16K static RAM, using an access time of 70 ns as the point of comparison, the CMOS chip dissipates approximately 130 mW while the NMOS chip dissipates 280 mW [8.1]. Low power operation means simplicity and low cost in packaging, both at chip level and system level. In a battery operation environment, the use of CMOS circuits is essential for minimum power consumption.

In the past, NMOS technology has enjoyed a simpler process as compared to CMOS. In a typical CMOS process, additional masks for the well, field, channel, and source/drain implants are necessary to fabricate both the n and p-channel transistors in the same circuit. But as the devices are scaled down in dimension, the device structures become much more complicated, and the process for NMOS becomes significantly more complex. Two channel implants instead of one, side-wall spacers, tip-implants in addition to the conventional source/drain implants, two-level metallization, and more advanced isolation techniques, are being employed in VLSI circuits. It becomes clear that for NMOS, the advantage of process simplicity over CMOS has been reduced significantly. The additional processing steps used in CMOS circuits are now only a small percentage of the total processing steps.

In the CMOS circuit, the p-channel transistor has lower performance compared with the n-channel transistor because of the lower mobility of holes (current carrier for p-channel) than of the electrons (current carrier for n-channel), with 200 versus 600 cm^2/V-sec. In the long channel device, the difference in current drive is about a factor of three. Thus CMOS circuits are traditionally slower than NMOS circuits. In order to have symmetry in the invertor transfer characteristics, it is necessary to make the p-channel transistor much wider than the n-channel to compensate for the mobility difference. This means a disadvantage in packing density compared with NMOS circuits. But as the channel lengths of the transistors are scaled down, this factor has become less important due to velocity saturation effects of the n-channel device.

As has been shown in Chapter 6, the current-voltage characteristics of MOSFETs for drain bias below saturation can be approximated by the simple equation:

$$I_{DS} = \frac{W}{L}\mu C_{ox}[(V_{GS} - V_T)V_{DS} - \frac{1}{2}V_{DS}^2] \qquad (8.1)$$

for long channel devices. It can be seen that the current drive is proportional to the mobility, and that the transconductance (g_m) in the saturation region is given by:

$$g_m \equiv \frac{\partial I_{DS}}{\partial V_{GS}} = \frac{W}{L}\mu C_{ox}(V_{GS} - V_T) \qquad (8.2)$$

We see that g_m is proportional to the mobility multiplied by the factor $(V_{GS}-V_T)$ when the device is in the saturation regime, and is inversely proportional to the channel length. But when velocity saturation occurs along the entire channel, the current-voltage characteristics are given by [8.2]:

$$I_{DSAT} = WC_{ox}V_s(V_{GS} - V_T) \qquad (8.3)$$

where I_{DS} is proportional to the saturation velocity (V_s). The corresponding g_m is a constant, and is independent of the channel length. Due to the higher mobility of the electrons, velocity saturation occurs at a lower electric field, therefore at longer channel length than for the holes, at the same drain bias. This means that as the channel lengths of n- and p-channel transistors are scaled down, the current drive of the n-channel device tends to saturate to a constant value independent of the channel length, while the p-channel device will gain in current drive until its current carriers exhibit the velocity saturation effect, at a shorter channel length. The comparison of n and p-channel transistor current drives measured as a function of L_{eff} is shown in Fig. 8.1. The current drive of the n-channel transistor begins to deviate from the $1/L_{eff}$ dependence due to velocity saturation, while the p-channel transistor current drive still obeys the $1/L_{eff}$ dependence. Because of this effect, the difference between n- and p-channel performance is reduced as the devices are scaled down.

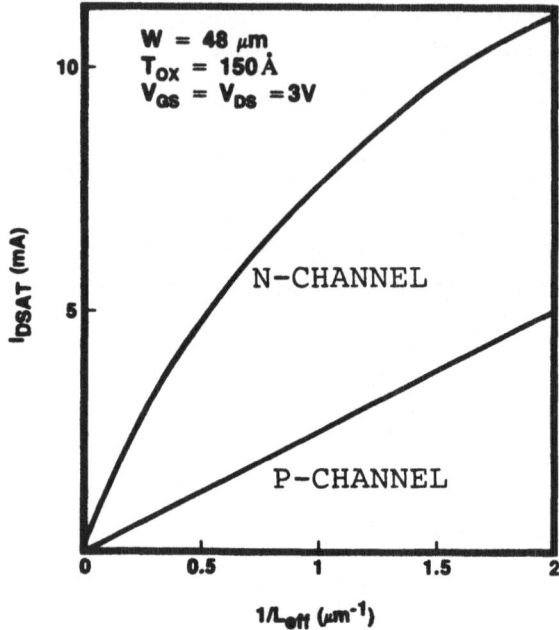

Fig. 8.1 Current drive vs. $1/L_{eff}$ for n and p-channel
          MOSFETs.

The density and speed of CMOS circuits can also be enhanced by creative circuit designs such as the domino-CMOS [8.3], where a small percentage of the devices are p-channel. In this case, the speed and density of the circuit is very close to NMOS circuits, while the power consumption is much reduced.

CMOS also has another major advantage, which is a very large noise margin. Unless major malfunction of the transistors has occurred, the circuit will function at least logically, although the power consumption and speed specification may be violated.

So far the advantages of CMOS circuits have been presented. However, there are also some major concerns in CMOS technology. The best known problem is latchup [8.4]-[8.5]. This subject will be covered in more detail in Chapter 9, which discusses the use of trench isolation to prevent latchup, and the problems that are associated with this technique. The other concerns of submicron CMOS technology are device issues, which are of a more general nature.

The design of short channel transistors is complicated by the effect of drain-induced barrier lowering (DIBL), as described in the previous chapter. Although the physics of DIBL in n-channel transistors has been discussed in that chapter, the DIBL that occurs in the p-channel transistor is somewhat more complicated. This will be discussed in detail in the next section. Other device concerns such as the drain to source avalanche breakdown of the n-channel transistors [8.6] (the p-channel transistor does not have this problem as severely as the n-channel because of the much lower impact ionization coefficient of the holes), as well as device degradation due to hot carrier effects [8.7] are of a general nature, not specific only to CMOS. In Chapter 11, the simulation of device structures used to reduce the hot carrier effects will be presented. In this chapter, the major emphasis will be on minimizing the residual leakage current of the p-channel transistor due to DIBL.

As will be shown later, the reason for the more severe DIBL problem in the p-channel transistor is because n^+ polysilicon is used for gate material. If p^+ polysilicon is used for the p-channel transistor, then the DIBL problem for the p-channel will be reduced to the same degree as for the n-channel transistor. But a combination of the n^+ and p^+ polysilicon will cause process complications. One of them is the processing of the p^+ polysilicon gate. Boron is used to dope the polysilicon to form the p^+ gate. But boron may diffuse through the gate oxide, causing threshold shift. Since the gate oxide thickness is being scaled down, typically to 15 nm in submicron devices, this problem will become even more serious in VLSI. Another problem is the connection of the n^+ and p^+ polysilicons within the chip. The obvious choice would be the use of silicides. This would require the development of a silicide process. Also, the n^+ and p^+ dopant may diffuse through the silicide grain boundaries to compensate each other. In this chapter, n^+ polysilicon gate is assumed for the process. It will be shown that by carefully choosing the device parameters, the leakage current can be minimized at effective channel lengths of 0.5 µm. For further reduction of the channel length, p^+ polysilicon or some other gate material having the appropriate work function must to be used.

8.2 Development of the Submicron P-Channel MOSFET Using Simulations

In this section, the development of the p-channel MOSFET will be presented. Extensive use has been made of simulations to fully understand the physics of the device. The relationship of the device structure to the DIBL effect, or the leakage current, is discussed. A guideline for the p-channel transistor structure which should reduce the leakage current will be presented.

P-Channel Transistor Structure

The p-channel transistor is fabricated on a n-type substrate or in an n-well in a p-type substrate. The fabrication procedure is similar to the n-channel transistor. The major difference in structure between the n- and p-channel is that the n-channel has a channel implant of the same impurity type as the substrate, while the p-channel implant is of the opposite type as the substrate. To understand better why this is necessary, let us consider the threshold voltage of an n-channel transistor with long channel length.

For an n^+ polysilicon gate over p-type silicon, assuming no interface charge and zero substrate bias, and a uniform bulk impurity concentration of N_B, the threshold voltage is given by:

$$V_T = \phi_{ms} + 2\psi_B + \frac{\sqrt{2\varepsilon q N_B(2\psi_B)}}{C_{ox}} \qquad (8.4)$$

where the first term is the work function difference between the n^+ polysilicon gate and the silicon substrate, ψ_B the potential difference between the intrinsic and hole Fermi level (~0.35 V), and the last term the voltage drop across the depletion region. The value for the work function difference is approximately -0.9 V and the term due to the depletion layer charge is about 0.1 V for $N_B = 1E15$ cm^{-3} and gate oxide thickness of 15 nm. Therefore, for a reasonable threshold voltage of typically 0.7 V, a surface doping concentration much higher than the typical substrate used in fabrication is needed. The reason that we want a low bulk doping concentration is to reduce the substrate bias effect and parasitic capacitance between the n^+ source/drain regions and the p-substrate.

For the p-channel transistor, the situation is different. In n-well CMOS, which will be assumed in this study, the n-well doping concentration is in the order of 1E16 cm^{-3} near the surface. This is necessary because one needs about an order of magnitude difference in impurity concentration between the n-well and the p-substrate, for process control reasons. Secondly, the n-well must have a high enough impurity concentration to avoid bulk punchthrough of the p-channel transistor, as well as to reduce latchup susceptibility by lowering the n-well resistivity (see Ch. 9). Substituting the n-well doping concentration value into the equation, the values of the three terms become -0.2, -0.7 and -0.35 respectively. It can be observed that the threshold voltage is about -1.2 V. Since a high threshold voltage means low current drive, the magnitude of this threshold voltage is too high for most applications.

In order to reduce the magnitude of the threshold voltage, a so called counter-doping technique is employed, which is an implantation of boron into the channel [8.8]. The dose is typically low, on the order of 5E11 cm^{-2}, and the profile is very shallow, typically 0.15 μm. The boron impurities are all depleted, and effectively acts like a layer of interface charge. Using this simplified argument, the threshold voltage as a function of the substrate doping and counter-doping can be estimated for long channel lengths. For comparison, Fig. 8.2 and 8.3 show calculations for the threshold voltages of both n and p-channel transistors with n^+ polysilicon gate. The n-channel calculations provide an idea of the dependence of the threshold voltage on the impurity concentration at the surface for two different oxide thicknesses. The p-channel calculations in Fig. 8.3 show the counter-doping dose (within the silicon substrate) that is necessary to provide a threshold voltage of -0.6 V, for different n-well surface doping concentrations. This is a good example which shows that often one can use very simple models to provide quick estimates of the parameter values, at the same time gaining some physical insight into the problem.

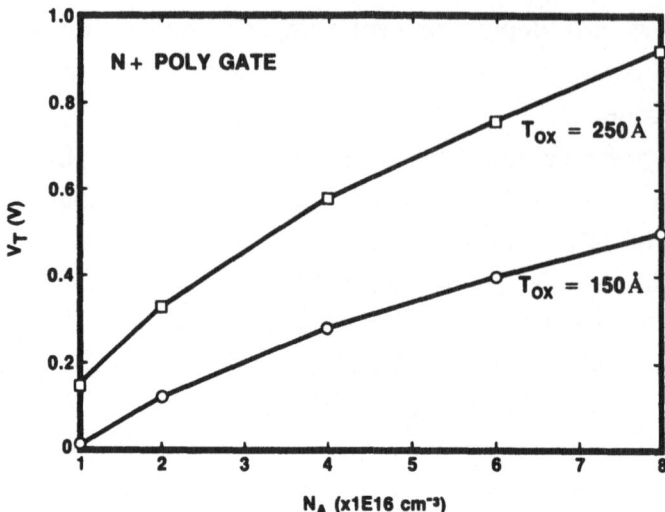

Fig. 8.2 Simulation of the n-channel threshold voltage
vs. p-type substrate doping concentration for
two oxide thicknesses.

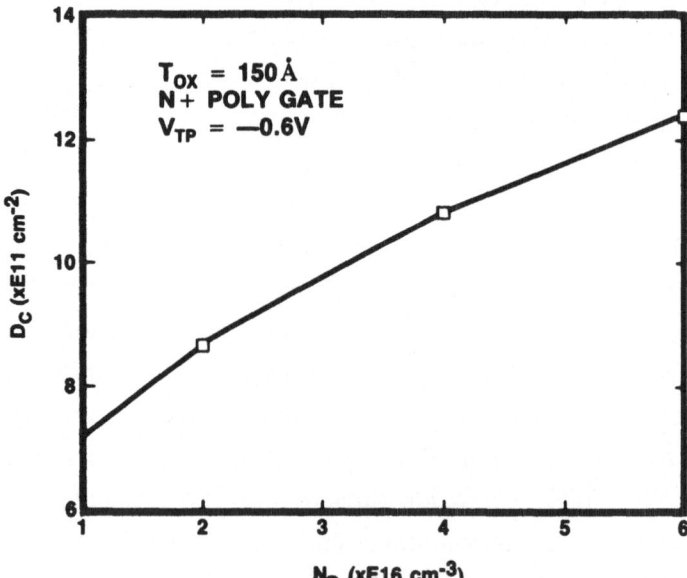

Fig. 8.3 Simulation of the counter-doping dose vs.
n-type substrate concentration for a fixed
p-channel threshold voltage of -0.6 V.

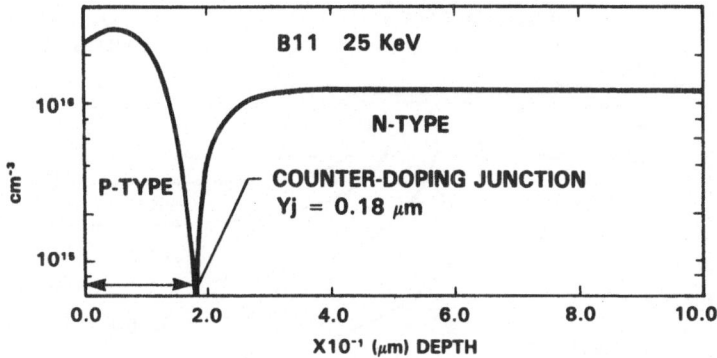

Fig. 8.4 Simulated channel profile for p-channel MOSFET.

To obtain a better understanding of the channel profile, the SUPREM program is used to simulate the channel profiles. Fig. 8.4 shows the simulated channel impurity profile of the p-channel transistor, versus distance from the silicon surface into the substrate. The counter-doping is performed by ion implantation of boron (B_{11}) at 25 KeV, with a dose of $7E11cm^{-2}$. Subsequent temperature cycles during the device fabrication give rise to a junction depth of about 0.18μm. The depth of the junction is also related to the fact that when B_{11} is implanted into silicon, there is channeling such that there is a long tail in the distribution going into the substrate [8.9]. If BF_2 is used instead, then the channeling effect can be reduced since the silicon surface is amorphized during the implant. This will be discussed later. Note that this implant profile is the exact opposite of what is in the n-channel transistor, as is shown in Chapter 6, Fig. 6.8. Because of this counter-doping, the effective substrate doping of the transistor can be thought of as being reduced, thus the drain electric field penetrates more towards the source, thereby giving rise to a higher susceptibility to DIBL.

Using the profile generated from the SUPREM program, the p-channel transistor structure can now be studied using the GEMINI program which solves for the potential profile within the device structure. Fig. 8.5 shows the p-channel transistor structure, with n^+ polysilicon gate, generated by the GEMINI program. On the left is the p^+ source and on the right is the p^+ drain. The counter-doping junction is indicated in the

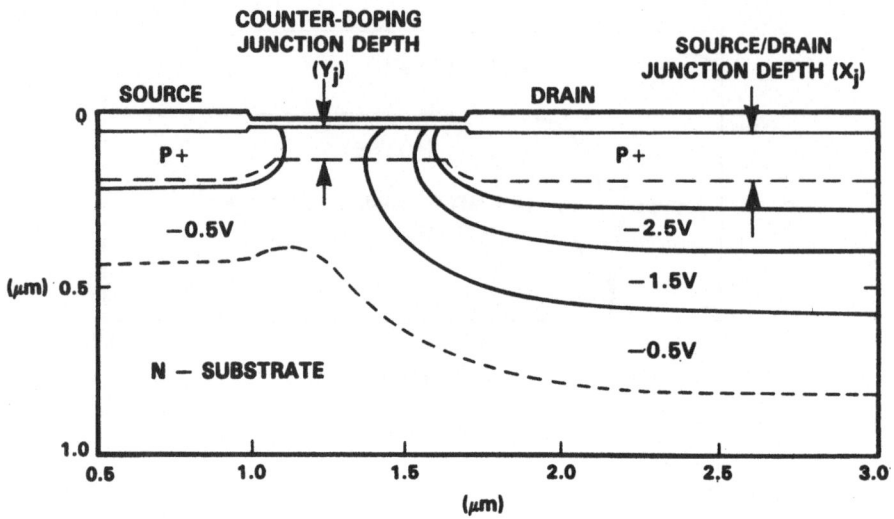

Fig. 8.5 Simulation of the p-channel MOSFET by GEMINI.

figure. The gate oxide thickness is 25 nm. For simplicity, the substrate doping concentration is assumed to be a constant and equal to 1E16 cm^{-3}. This is not exactly true in the n-well CMOS process, since the profile is really a Gaussian profile. But the correction to the device characteristics is small, since the device characteristics are mainly dependent on the surface concentration, down to about 0.5μm. The bias voltage on the gate, source, drain, and substrate are 0, 0, -3 and 0 V respectively. The two-dimensional potential contours are generated by the program.

Drain-Induced Barrier Lowering in P-channel Transistors

This section describes the study of the DIBL problem in p-channel transistors with n$^+$ polysilicon gate, using simulations. The technique of simulating the p-channel transistor structure has been presented in the last section. Here the details of the potential distribution within the device structure are described.

The DIBL effect is dependent on three critical device parameters. They are: the gate oxide thickness (T_{ox}); counter-doping junction depth (Y_j); and source/drain junction depth (X_j), for a fixed channel length. In this discussion, the effective channel length (L_{eff}) is fixed at 0.5μm. The DIBL effect is characterized by the residual leakage current (I_L), which is the current at a drain bias of -3 V and gate bias at 0 V. Another parameter is the subthreshold slope (S), which degrades with increasing DIBL. The DIBL effect can also be characterized by the sensitivity of the threshold voltage on the channel length, when the drain is biased at -3V. The leakage current, subthreshold slope, and threshold variations are simulated as functions of the device parameters. The L_{eff} is 0.5 μm and T_{ox} is 25 nm for all simulations unless stated otherwise.

P-channel transistors with different values of Y_j (0.1, 0.15, 0.2 μm) but fixed values of threshold voltage of about -0.65 V are simulated. Since the DIBL effect is very sensitive to the threshold voltage, (the higher the magnitude of the threshold voltage, the less is the DIBL effect due to lower counter-doping dose or higher n-well doping concentration) the amount of counter-doping dose has been adjusted so that the threshold voltage is similar for the three cases to give a valid comparison. The potential profiles are shown in Fig. 8.6 for the three cases, with Y_j values increases from top to bottom. From the potential contours, it can be observed that as Y_j is increased, the potential spreads more from the drain to the source. This indicates more DIBL effect. The major punchthrough path can be either at the surface or in the bulk. The bulk punchthrough path can be traced out by joining the points of the potential contours which are most extreme towards the source, as shown in the last figure of Fig. 8.6. The bulk punchthrough path is within the counter-doping junction, indicating that the counter-doping junction is enhancing the DIBL effect. Note that this is different from the case of n-channel transistors where the bulk punchthrough path under a large drain bias is quite deep (~0.7 μm) below the surface, due to the high surface impurity concentration and lightly doped substrate (Fig. 7.6 in Chapter 7).

Fig. 8.7 shows the potential profile as a function of the vertical distance from the silicon surface into the bulk, at a horizontal distance of 0.2 μm away from the source. This horizontal position is chosen because it is near the "saddle point" (see Ch. 7) of the potential distribution, hence

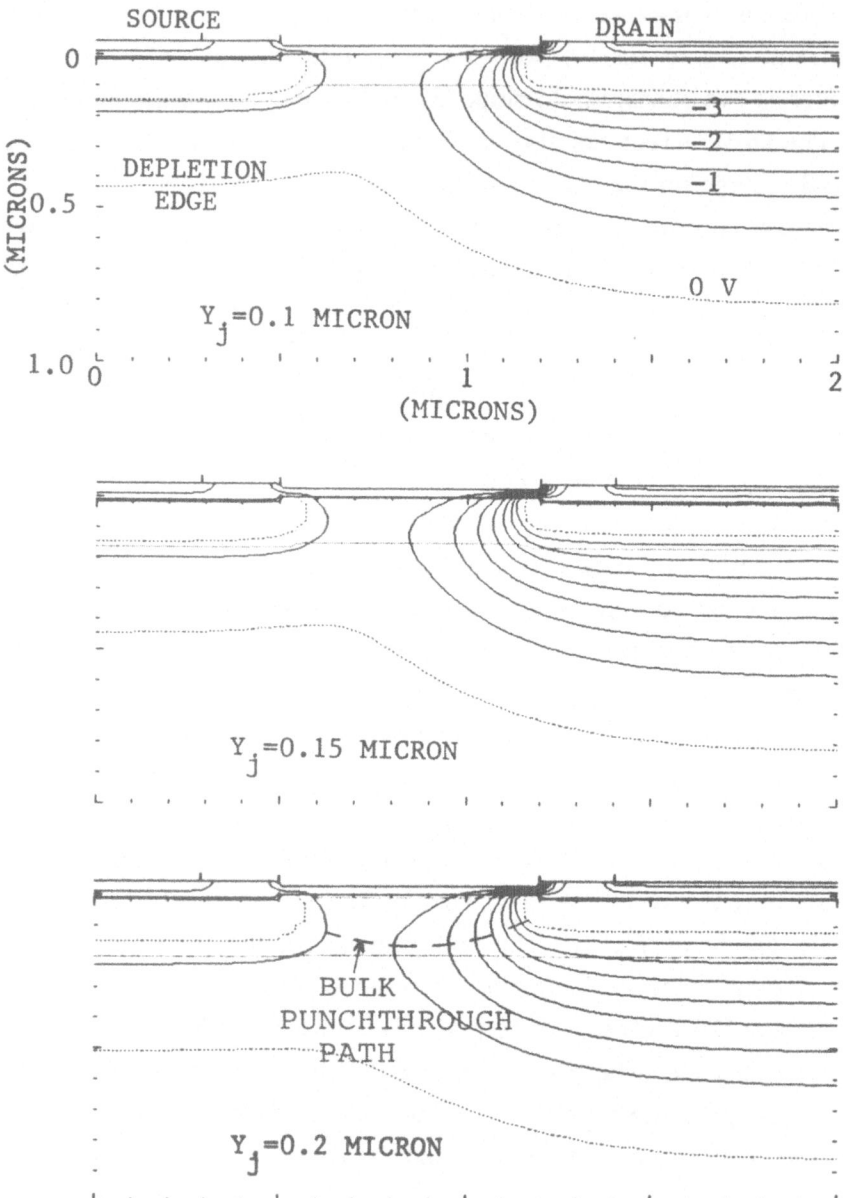

Fig. 8.6 Simulated 2-D potential profile for p-channel
MOSFET with $Y_j = 0.1$, 0.15, and 0.2 μm.
The drain, gate bias is -3 and 0 V
respectively.

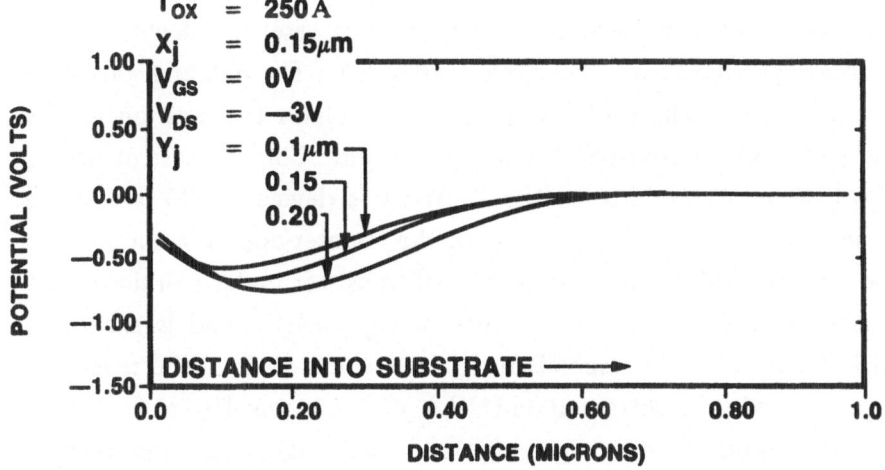

Fig. 8.7 Channel potential profile vs. vertical distance for a p-channel MOSFET with L_{eff} of 0.5 μm, with three different Y_j values.

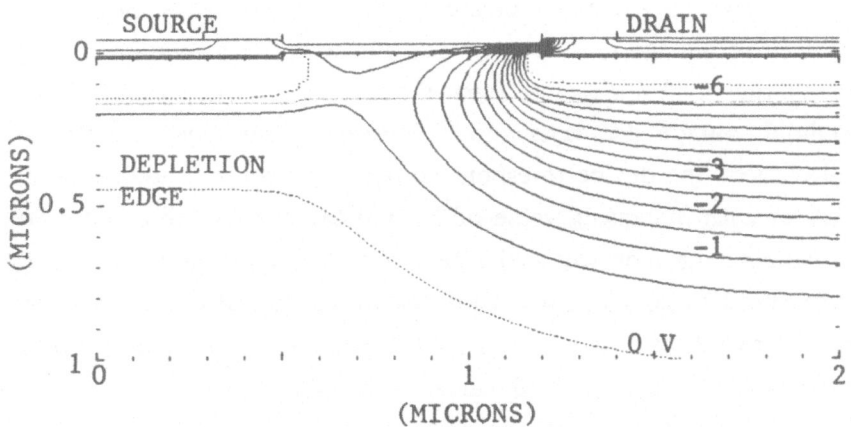

Fig. 8.8 2-D potential profile for a p-channel MOSFET with L_{eff} of 0.5 μ, drain and gate bias of -6 and 0 V respectively.

determining the punchthrough current path. The potential (and hence also the potential barrier) in the bulk is lower for higher values of Y_j.

To further investigate the DIBL effect, the device is simulated for the condition where the drain to source leakage current at zero gate bias is 100 nA, for a device width of 50 μm. In this case, the device is in a "punchthrough" condition. The drain bias is -6 V. The potential profile is shown in Fig. 8.8. The GEMINI program calculates the position of the saddle point, which controls the leakage current. For this case, it occurs at $X = 1.18$ and $Y = 0.14$ μm. The Y_j for this device is 0.15 μm. Note that the saddle point is near the edge of the counter-doping junction.

The subthreshold slope of the p-channel transistor under a drain bias of -3 V is simulated for different values of X_j and Y_j, and is shown in Fig. 8.9. Each curve corresponds to a fixed value of Y_j. The threshold voltage, defined by a current $100nA*W_{eff}/L_{eff}$, is specified at each data point. The simulated data show that the subthreshold slope is strongly dependent upon X_j and Y_j.

The interpretation of the graph is as follows.   With a value of $X_j = 0.1$ μm, devices with different values of Y_j (0.1, 0.15, 0.2 μm) are simulated, and the counter-doping doses are adjusted to give approximately the same threshold voltage (V_T). The data show that S is very sensitive to Y_j, and increases with Y_j. The values of X_j is then increased to see the sensitivity of the device performance on X_j. It can be seen that as X_j increases from 0.1 to 0.25 μm, the magnitude of the threshold voltage decreases and S increases, both due to DIBL.   The sensitivity of the device threshold voltage and subthreshold slope to X_j increases with increasing value of Y_j, indicating that the counter-doping is enhancing the short channel effect. For example, for $Y_j = 0.1$ μm, the subthreshold slope increases from 104 to 122 mV/dec for X_j increases from 0.1 to 0.25 μm; while for $Y_j = 0.2$ μm, the slope increases from 140 to 194 mV/dec. For a fixed value of $X_j = 0.15$ μm, $V_T = -0.67$ V, and $T_{ox} = 25$ nm and 15 nm, the dependence of S on Y_j is shown in Fig. 8.10. For $T_{ox} = 25$ nm, S increases from 112 to 170 mV/dec, when Y_j increases from 0.1 to 0.2 μm, showing the effect of Y_j on the DIBL. The subthreshold slope improves for a thinner gate oxide of 15 nm, but the

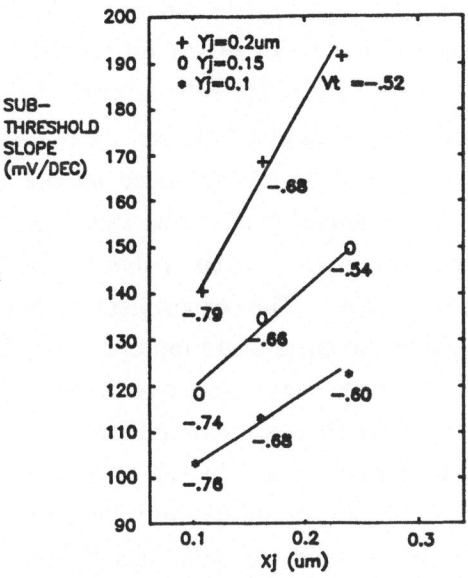

Fig. 8.9 Simulated subthreshold slope vs. X_j and Y_j for p-channel MOSFET. $L_{eff} = 0.5$ μm, $T_{ox} = 25$ nm, $V_{DS} = -3$ V. Vt is the threshold voltage.

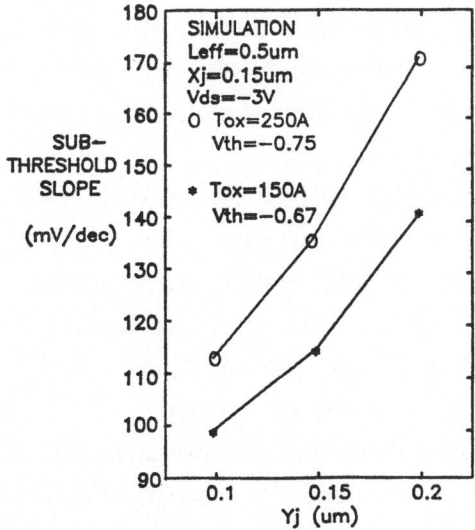

Fig. 8.10 Simulated subthreshold slope vs. Y_j and T_{ox}.

dependence of S on Y_j is only slightly reduced. The figure shows that the gate oxide thickness has a major effect on the subthreshold slope, as expected. It can be seen here that for submicron CMOS, thin gate oxide is essential for controlling the subthreshold leakage [8.10],[8.11].

The DIBL effect can also be characterized by the sensitivity of the threshold voltage to the channel length. The reduction of the threshold voltage as the channel length is reduced is due to the potential barrier lowering by the drain bias. The experimental data for p-channel MOSFET with gate oxide thickness of 15 nm are shown in Fig. 8.11. This reduction in threshold voltage may cause circuit performance problems, especially in analog design such as sense amplifiers in memory circuits. It may also cause excessive off current in the transistors, yielding a large standby power. In general, one would like to reduce the sensitivity of the threshold voltage to the effective channel length, thus minimizing the threshold voltage variations throughout the circuit. Fig. 8.12 shows the simulation of this sensitivity for the p-channel transistor with two values of Y_j. It can be seen that the sensitivity increases with increasing Y_j, which implies that Y_j should be minimized as much as possible. The sensitivity will also be reduced if a thinner oxide is used. This is because the channel potential will be more controlled by the gate, thus reducing the drain bias effect. The threshold sensitivity for $T_{ox} = 25$ and 15 nm are shown in Fig. 8.13. The results show that thin oxides are essential for very short channel devices.

The leakage current (I_L), defined as the drain to source current at zero gate bias, is simulated for different values of X_j and Y_j. The drain to source bias is -3 V. The results are shown in Fig. 8.14. The interpretation of the graph is the same as for Fig. 8.9. The leakage current is calculated at $X_j = 0.1$ μm for $Y_j = 0.1$, 0.15, and 0.2 μm, with approximately the same threshold voltage. As the value of X_j is increased, the leakage current increases exponentially. Also, for a fixed X_j (for example, 0.15 μm), and for approximately the same V_T, the leakage current increases exponentially with Y_j. The simulated data show that in order for the leakage current to be below 1 nA for a channel length of 0.5 μm and width of 50 μm (this criterion is arbitrary,

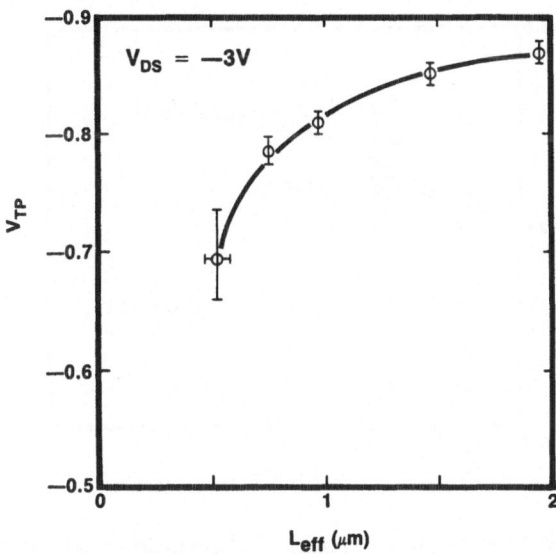

Fig. 8.11 Experimental data of p-channel threshold voltage vs. L_{eff}, for gate oxide thickness of 15 nm.

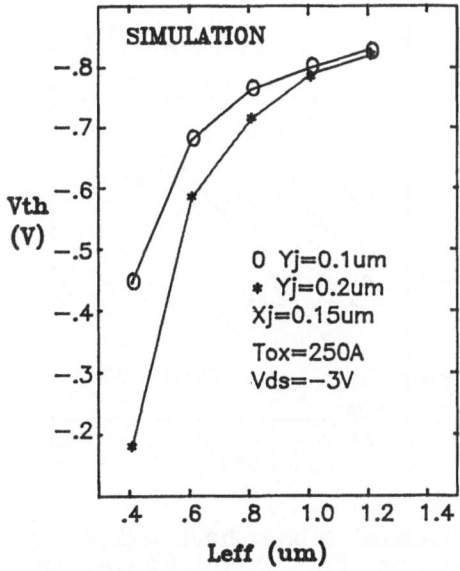

Fig. 8.12 Simulation of threshold voltage vs. L_{eff} for p-channel MOSFET.

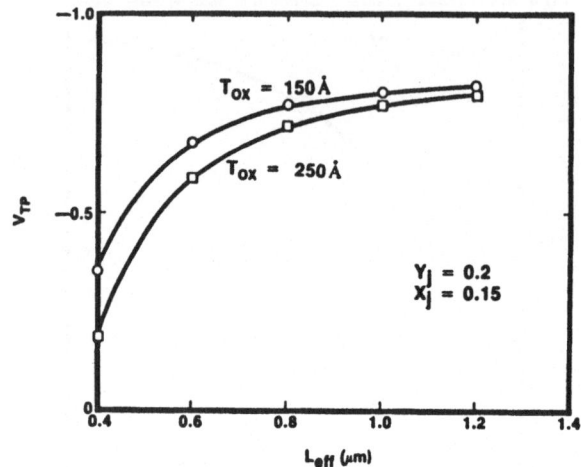

Fig. 8.13 Simulated p-channel threshold voltage vs. L_{eff} for two different gate oxide thicknesses.

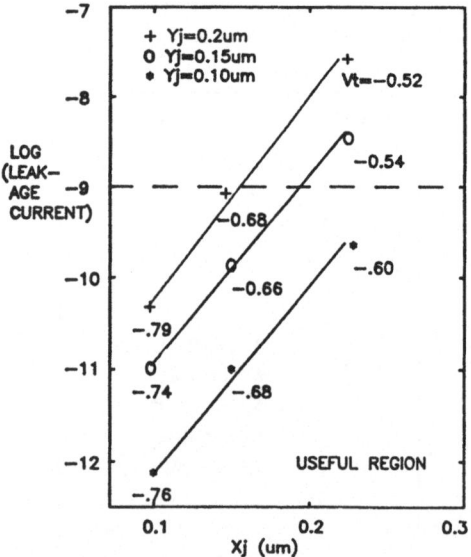

Fig. 8.14 Simulated subthreshold leakage current vs. Y_j and X_j. $L_{eff} = 0.5$ μm, $W = 50$ μm, $T_{ox} = 25$ nm, $V_{GS} = 0$V, $V_{DS} = -3$V.

and is really dependent on the application), X_j and Y_j must be below 0.2 and 0.15 μm respectively. If Y_j can be maintained at 0.1 μm, then the criterion for X_j is relaxed to 0.25 μm [8.8]. This simulation is very useful in guiding the development of the submicron CMOS process. It provides the limiting values of the critical parameters such as T_{ox}, X_j and Y_j. These parameters have significant effect on the processing procedures. For example, if a very shallow source/drain junction is required, this will put a serious constraint on the temperature cycles of the subsequent backend process, such as interlevel dielectric reflow or planarization. It is always necessary to have estimates on the limiting values of these device parameters during process development.

Arsenic Implant Technique to Reduce Y_j

The conventional approach to the counter-doping process for adjusting the p-channel threshold voltage is the implantation of B_{11} at 25 KeV, which produces a junction depth of about 0.18 μm. The simulated profile is shown in Fig. 8.4 in section 8.2. As has been shown, this counter-doping causes severe short channel effects such as subthreshold leakage. A new process to produce a shallow counter-doping junction depth has been developed with the help of simulations. This technique uses an arsenic implant immediately following a BF$_2$ implant. BF$_2$ is used because it reduces the channeling effect that is observed in the case of B_{11}. This produces a shallower junction compared with B_{11}. Furthermore, arsenic diffuses relatively slowly during process temperature cycles. It forms a steep profile at the counter-doping junction, and maintains a shallow junction depth. The simulation is shown in Fig. 8.15, and gives a junction depth of 0.09 μm.

One major concern is whether this arsenic implant will introduce a much higher n-type impurity concentration near the surface, which will increase the capacitance between the p^+ diffusion islands and the substrate. As can be seen in the simulation, the n-type impurity concentration is only slightly higher than the conventional technique. Also, the position of the depletion edge under the p^+ diffusion will be beyond the peak of the arsenic profile which only extends to about

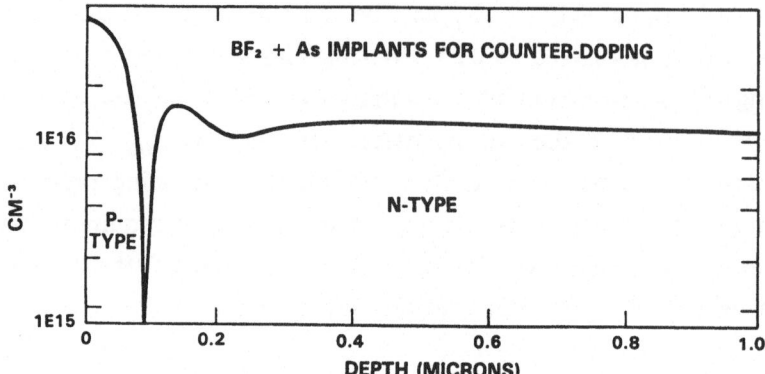

Fig. 8.15 Simulated channel profile for p-channel with
BF$_2$ and As counter-doping.

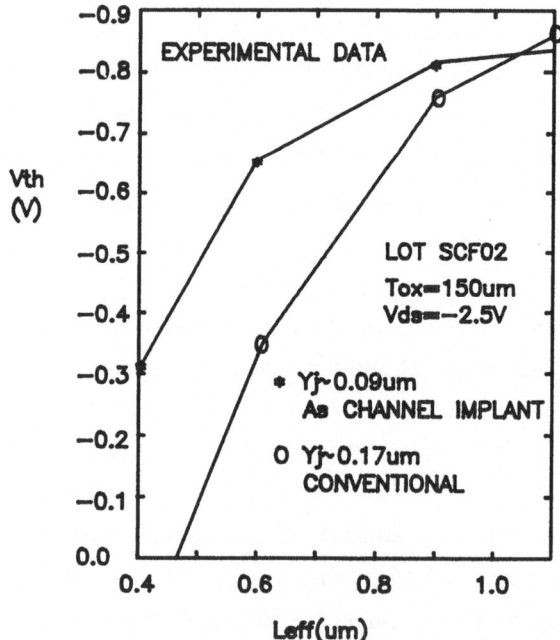

Fig. 8.16 Experimental data of threshold voltage vs.
L_{eff} for p-channel with different Y_j.

0.2 μm. It is clear that this technique will not significantly increase the diffusion parasitic capacitance. One disadvantage of this technique is the process variation sensitivity. The boron channel implant is compensated by an arsenic implant, which requires good process control. Another disadvantage is that this technique requires an extra masking step to prevent the arsenic implant from going into the n-channel transistors. The n-well mask can be used and the alignment is non-critical.

Submicron p-channel transistors with reduced short channel effects were fabricated using this technique. Fig. 8.16 shows the experimental data of the threshold voltage vs. effective channel length, for a drain to source bias of -2.5 V. The devices with larger values of Y_j showed more threshold roll-off as the channel length is reduced.

8.3 N-Channel Transistor Simulations

The design of submicron n-channel transistors is easier compared with the p-channel transistor in terms of threshold and DIBL control. The n-channel transistor with n^+ polysilicon gate requires a channel implant of the same type as the substrate, which reduces the short channel effects. Two-dimensional simulations are used to calculate the optimal channel implant profiles. There are other device issues that are more specific to n-channel transistors, which are also discussed in the following sections.

Concerns for Submicron N-Channel MOSFETs

There are two major device issues specific to n-channel transistors, in addition to the threshold voltage and DIBL control which are of general concern for submicron devices. They are the avalanche breakdown between the source and drain [8.6] and device performance degradation due to hot electrons [8.7]. The n-channel device is more susceptible to avalanche breakdown than the p-channel, because the impact ionization coefficient for electrons is much higher than that of the holes. Also, for conventional source/drain structures, the n-channel transistor source and drain have more abrupt junctions, which means a higher electric field than the p^+ junctions in the p-channel transistor. The higher ionization coefficient and the higher electric field cause a higher level of substrate current. This current produces a voltage drop across the substrate, which reduces the potential barrier between the source and substrate. Thus the channel current increases and produces more substrate current. This produces a positive feedback action which eventually causes the avalanche breakdown of the device.

Degradation due to hot carrier effects is more serious for the n-channel transistor, because of the higher drain electric field and higher mobility of the electrons. Also, the potential barrier between the gate oxide and the channel is higher for the holes, making the hole injection into the oxide less likely than electron injection. Thus the understanding of the electric field distribution at the drain region as a function of the drain structure is very important. New device structures such as the Lightly-Doped Drain (LDD) structure have been developed to reduce the electric field. The electric field can be simulated by the CADDET program. This will be presented in detail in Chapter 11.

Threshold and DIBL Control

The threshold of the n-channel transistor as a function of the channel profile can be simulated by SUPREM for the long channel, and by the combination of SUPREM and GEMINI for the short channel transistors. The channel profile will depend on the application of the devices, as well as the structural parameters such as gate oxide thickness. For example, the amount of doping will be less if negative substrate bias is used in the circuit. In the discussion below, the substrate bias is assumed to be zero, as is the case in most CMOS circuits.

From the discussion in the previous section, it can be seen that for p-channel transistors with channel length of 0.5 μm, a thin gate oxide of 15 nm must be used to reduce the residual leakage current. Hence the channel implant for the n-channel transistor will be based on this gate oxide thickness. Fig. 8.17 shows the channel profile which will produce a threshold voltage of 0.5 V for long channel transistors. But the threshold voltage will be reduced when the channel length is reduced to 0.5 μm, and when under a high drain bias. This is studied by coupling the

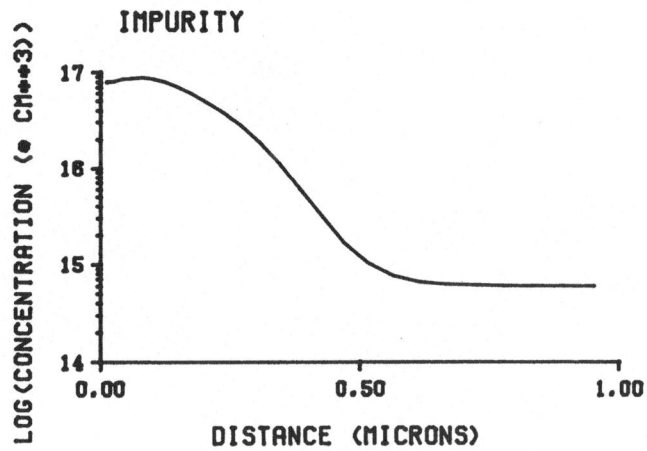

Fig. 8.17 N-channel impurity profile for submicron devices.

channel profile with the GEMINI program, and specifying the desired drain bias. Fig. 8.18 shows the results of such a simulation. The DIBL effect is evident from the shape of the potential profile which extends from the drain towards the source. The high impurity surface concentration has prevented the punchthrough problem near the surface. This reduces the leakage current under a drain bias of about 3 V, where the potential barrier minimum point is found to be at the surface. At higher drain bias, the punchthrough current path will be through the more lightly doped bulk. The reason for this has been explained in Chapter 7.

The threshold voltage as a function of the effective channel length is simulated for the case of 15 nm gate oxide and a drain bias of 3 V, using the GEMINI program. The result is shown in Fig. 8.19. The threshold roll-off is quite significant at L_{eff} of 0.5 μm, but is still tolerable for most applications. The roll-off can be reduced by using a higher channel implant dose, at the expanse of lower current drive, higher junction capacitance and body effect (see Table 6.1 in Chapter 6).

Source/Drain Structures

The source/drain structure of submicron devices are determined by

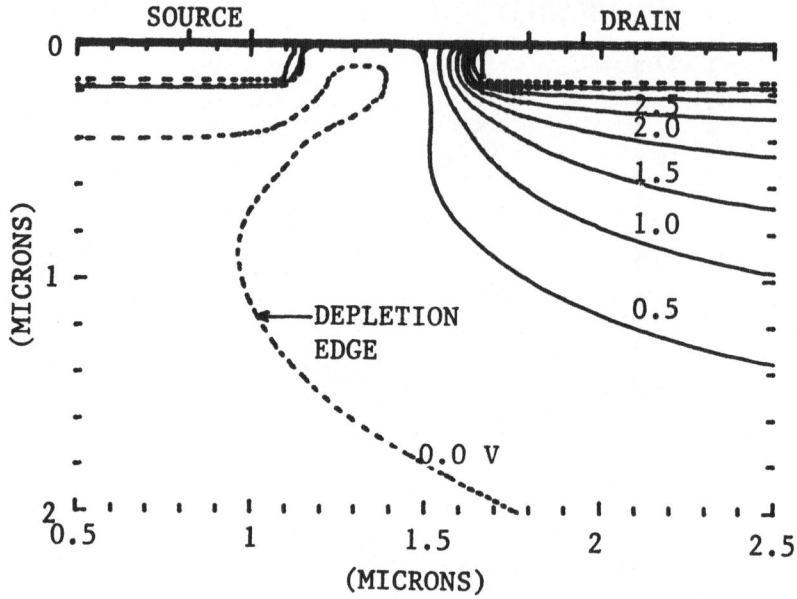

Fig. 8.18 2-D potential profile of submicron n-channel
MOSFET.

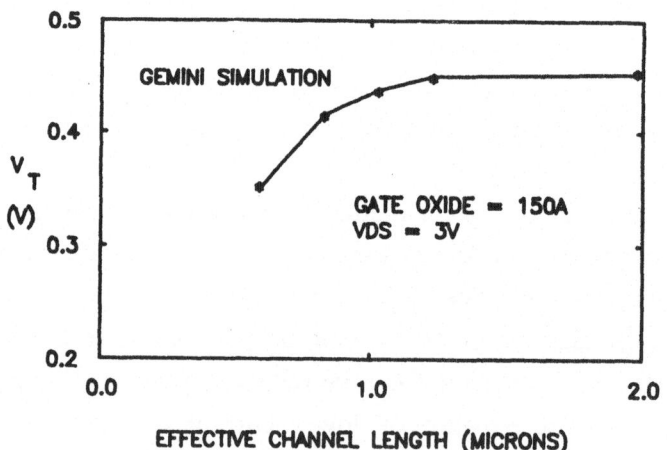

Fig. 8.19 Simulated threshold voltage vs. L_{eff} for
n-channel MOSFET.

several considerations. The first is the high electric field near the drain. This causes hot carrier degradation problems as well as avalanche breakdown, as mentioned previously. Two types of drain structures are used for reducing the electric field. They are the LDD [8.12] and the double diffused junction [8.13] structures. The double diffused drain structure uses a combination of arsenic and phosphorous implant to form a more graded junction, thus reducing the electric field at the junction. The LDD structure is formed by using a low dose tip implant before the formation of the side-wall spacer, followed by the formation of the spacer and the n$^+$ diffusion region. Fig. 8.20 shows the simulation of

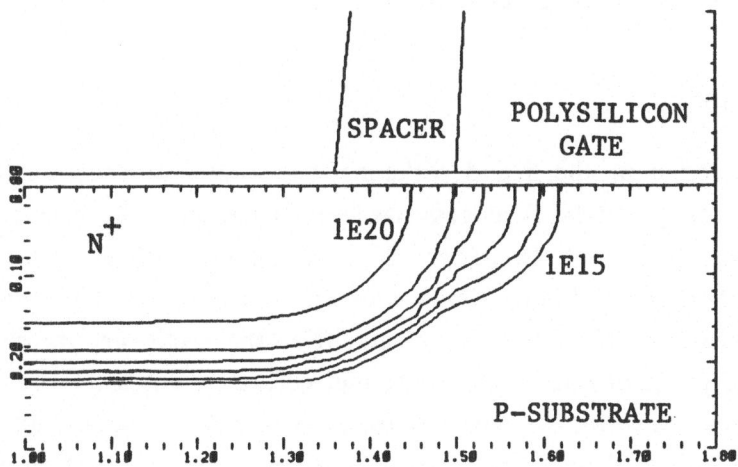

Fig. 8.20 Simulation of the LDD structure by SUPRA.

such a structure using the SUPRA program. The tip implant reduces the electric field because of the light doping. The detailed analysis of the LDD structure and its effect on the drain electric field will be presented in Chapter 11.

The second consideration in the source/drain structure is process control. For submicron gate length, side-wall spacers are commonly used. This means that either a tip implant is used, or the source/drain region has to be diffused far enough into the channel region to ensure overlap between the gate and the n$^+$ regions. In this respect, the tip implant is more desirable since the overlap is guaranteed, independent of side-wall spacer thickness variations. Also, the tip junction depth is only about 0.1 μm, which reduces the punchthrough problem at short channel lengths, for the same amount of gate to source/drain overlap.

The third consideration is the process complexity. In NMOS technology, the tip implant does not require additional masking. But for CMOS, the tip implant requires an additional mask since the p-channel device has to be masked during the tip implant. The source/drain structure of the submicron n-channel transistor has to be determined by the application of the device. In any case, the conventional arsenic structure will not be adequate due to reliability problems. Either double diffused junction or LDD has to be used.

8.4 Summary

The design of the n and p-channel transistors for the submicron n+ polysilicon gate CMOS process has been presented. The major concern for the p-channel transistor is subthreshold leakage due to the counter-doping. Using simulations, the device can be designed to minimize the short channel effects. For the n-channel transistor, the leakage problem due to short channel effects is not as severe. But the n-channel transistor has more reliability concerns such as the hot electron effects and avalanche breakdown between the source and drain. This requires more complicated source/drain structures such as graded junction and LDD structures. The study of the source/drain structure becomes more important and Chapter 11 will deal with this subject in detail using simulations. Computer-aided design has been shown to be essential in advanced device development where device structures are becoming more and more complicated due to many effects related to short channel devices.

References

[8.1] F. Lee, N. Godinho, and C. P. Chiu, "Cool-Running 16K RAM Rivals

N-Channel MOS Performance," *Electronics*, Oct 6, 1981, pp. 120-123.

[8.2] Y. El-Mansy, "MOS Device and Technology Constraints in VLSI," *IEEE Trans. on Electron Devices*, <u>ED-29</u>, Apr 1982, pp. 567-573.

[8.3] L. C. Thomas and J. J. Molinelli, "VLSI Logic," *Tech Digest of ISSCC 1981*, pp. 230-231.

[8.4] R. R. Troutman, "Recent Developments in CMOS Latchup," *Tech. Digest of IEDM 1984*, pp. 296-299.

[8.5] D. B. Estreich, "The Physics and Modeling of Latch-up in CMOS Integrated Circuits," Stanford Electronics Labs Report #G-201-9, 1980.

[8.6] S. E. Laux and F. H. Gaensslen, "A Study of Avalanche Breakdown in Scaled N-Channel MOSFETs," *Tech. Digest of IEDM 1984*, pp. 84-86.

[8.7] C. Hu, "Hot Electron Effects in MOSFETs," *Tech. Digest of IEDM 1983*, pp. 176-181.

[8.8] K. M. Cham and S. Y. Chiang, "Device Design for the Submicrometer p-Channel FET with n+ Polysilicon Gate," *IEEE Trans. on Electron Devices*, <u>ED-31</u>, July 1984, pp. 964-968.

[8.9] D. V. Morgan, Ed., *Channeling: Theory, Observation and Applications*, New York:Wiley, 1973.

[8.10] T. Kobayashi, S. Horiguchi and K. Kiuchi, "Deep-Submicron MOSFET Characteristics with 5nm Gate Oxide," *Tech. Digest of IEDM 1984*, pp. 414-417.

[8.11] J. R. Brews, W. Fichtner, E. H. Nicollian, and S. M. Sze, "Generalized Guide for MOSFET Miniaturization," *IEEE Electron Devices Lett.*, <u>EDL-1</u>, Jan 1980, pp. 2-4.

[8.12] H. Katto, K. Okuyama, S. Megure, R. Nagai and S. Ikeda, "Hot Carrier Degradation Modes and Optimization of LDD MOSFETs," *Tech. Digest of IEDM 1984*, pp. 774-777.

[8.13] K. Balasubramanyam, et al, "Characterization of As-P Double Diffused Drain Structure," *Tech. Digest of IEDM 1984*, pp. 782-785.

Chapter 9

The Surface Inversion Problem in Trench Isolated CMOS

9.1 Introduction to Trench Isolation in CMOS

Trench isolation has recently been proposed for advanced CMOS

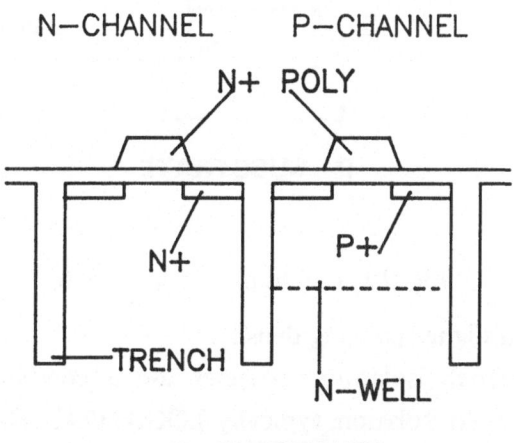

Fig. 9.1 Trench isolation CMOS technology.

processes [9.1]-[9.4]. Fig. 9.1 shows a trench isolated CMOS structure. The major advantage of trench isolation in CMOS is to reduce the latchup problem [9.5],[9.6]. Fig. 9.2 shows the latchup path for a CMOS structure, where the parasitic bipolar transistors are shown to form a positive feedback path. If there is excess current flow in the n-well and substrate, and if the substrate and n-well resistances are large enough, then significant voltage drops will occur in the substrate and n-well. Under this condition, the parasitic bipolar transistors can be turned on [9.5]. This phenomenon has been studied by many researchers in CMOS technology. The trench has been suggested as a good candidate for isolation between n-channel and p-channel transistors. For an n-well process it is expected that the lateral n-p-n parasitic bipolar transistor gain will be reduced, thereby increasing the latchup initiating current. In a p-well process, the reduction of latchup sensitivity is achieved from reduction of the lateral parasitic p-n-p current gain. Besides providing higher resistance to latchup, the trench also reduces the lateral side diffusion of the well,

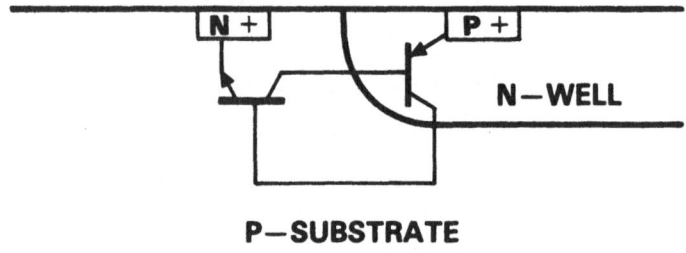

Fig. 9.2 Latchup path in CMOS.

hence allowing a higher packing density.

At present, trench isolation processes use a combination of trench isolation and a field isolation, typically LOCOS [9.4]. This is necessary to avoid leakage problems at the trench surface due to inversion, and also provide field area for interconnections. For our discussion, we assume that the trench is etched with vertical walls and refilled with oxide in a processing sequence that is consistent with low defect VLSI technology. It is interesting to see if the transistors can be placed adjacent to the trench side-wall. If this is possible, then it will provide the highest packing density for bulk CMOS technology, if the interconnection process is also optimized. Also, the narrow width effects on the transistors (see Chapter 10) are eliminated because the trench has a vertical wall, in contrast to the conventional LOCOS process. For LOCOS, the bird's beak has caused the loss of channel width in narrow width transistors, and also causes the increase of threshold voltage with decreasing channel width. Since trench isolation has no field implant, there is no boron encroachment into the active area.

It would be possible to place the transistors next to the trench if the trench surface is ideal. However, this may not be the case. The trench surface may be inverted due to positive charges at the trench surface or in the trench refill material [9.7],[9.8]. The effect of the plasma used to etch the trench may enhance the charge density at the trench surface. This problem is more serious for n-well CMOS because the n-well is biased at a positive voltage. The n-well in this case acts like an electrode which tends to invert the p-type substrate region opposite to the n-well. Also, the substrate doping concentration decreases with increasing distances into the substrate. This means that a relatively low charge density at the surface of the trench may invert the trench surface. If the trench surface is inverted, then there will be a conducting path connecting the n-well and the n$^+$ region. This increases the effective area of the collector of the lateral npn parasitic bipolar transistor, hence causing latchup problems. The inversion will also cause leakage between isolated n$^+$ regions and between the source and drain of the n-channel transistor. The problem of trench surface inversion has been studied using simulations to identify the problem under different processing and bias conditions. From the simulation results, appropriate application of the trench isolation is recommended. Experimental results are also presented.

9.2 Simulation Techniques

Before doing the numerical simulations, it is useful to roughly estimate the range of fixed charge density (Q_{SS}) which would cause inversion at the trench surface. A simple model would be an MOS capacitor with the n-well as the gate, the trench as the dielectric with the dielectric thickness equal to the trench width. Positive charge with density of Q_{SS} cm^{-2} is assumed to be at the trench surface. Fig. 9.3 shows the structure. Then the threshold voltage is simply given by equation [9.9]:

$$V_T = \phi_{ms} + \frac{Q_B - Q_{ss}}{C_{ox}} + 2\psi_B \qquad (9.1)$$

where the first term is due to the work function difference between the n-well and p-substrate, the second term due to the depletion charge density at inversion (Q_B) and trench surface charge density, and the last term is the surface potential bending necessary to achieve inversion. Here C_{ox} is the dielectric capacitance between the trench capacitor. The values of ϕ_{ms}, Q_B, and ψ_B can be easily calculated from the doping concentration of the n-well and p-substrate.

The threshold as a function of Q_{SS} for a p-substrate doping of 1E15 and 1E16 cm^{-3} and n-well doping of 1E16 cm^{-3} is shown in Fig. 9.4. It is clear that Q_{SS} is a problem for trench isolation for lightly doped substrate.

In order to investigate this problem in more detail, taking into account the effect of a varying p-type impurity distribution along the trench surface, two-dimensional simulations are performed. Also, the simulations can provide information about the region where inversion is most likely to occur.

The trench structure can be generated using the GEMINI program which solves Poisson's equation. The only caution in using this program is that the quasi-Fermi level along the channel is set equal to the drain quasi-Fermi level, even after inversion. This means that the results are valid only at or before inversion occurs at the trench surface. After inversion has occurred, the potential value at the trench surface would be overestimated, and so would be the extension of the inverted region.

The structure is generated using the MOSFET structure, with the thickness of the gate oxide equal to the trench depth. Then the source

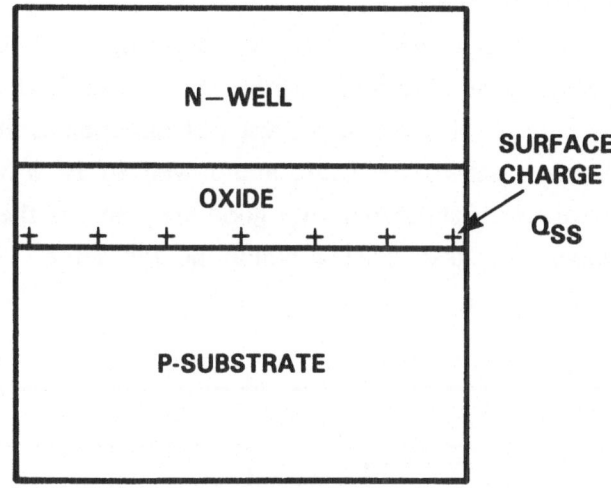

Fig. 9.3 Capacitor for trench surface inversion calculation. The oxide thickness is equal to the trench width.

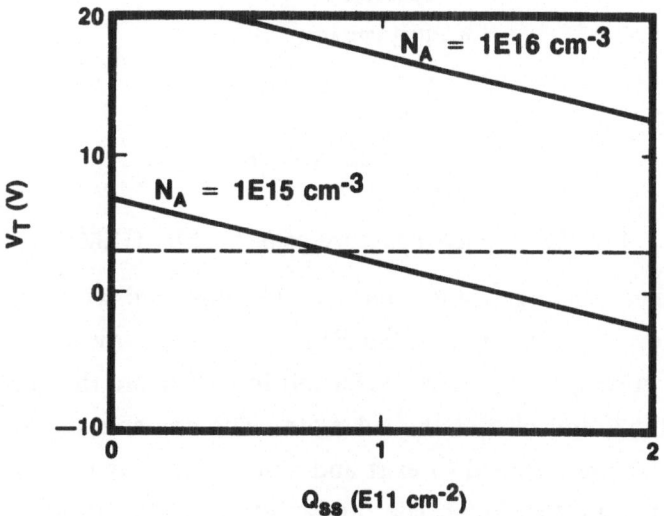

Fig. 9.4 Trench surface threshold voltage vs. Q_{SS} and p-type substrate concentration. The dotted line indicates a threshold voltage of 3 V.

and drain regions are raised to the top of the trench. The drain of the MOSFET becomes the n-well, and the source becomes an n⁺ region. The impurity profile can either be generated by the SUPREM program, or can be specified as Gaussian profiles in the GEMINI input file. The maximum number of grids in the vertical direction is 99, therefore, the trench depth is set to be 5 μm, and n-well to be 3 μm, such that a realistic structure is simulated with good accuracy. If the trench is made much deeper, the grid spacing would be too large. The number of

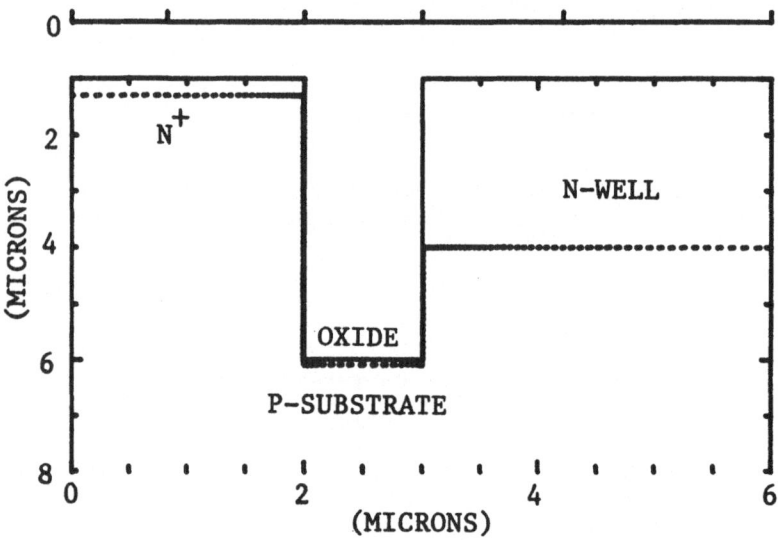

Fig. 9.5 Trench structure generated by GEMINI.

iterations for convergence is typically 300, compared with values of 100 or less for conventional MOSFETs.   Fig. 9.5 shows the generated structure. Also specified in the GEMINI input file are the applied bias at the n-well, n⁺, and also at the substrate. The contacts to the n⁺, n-well and substrate are assumed to exist and will not be shown. The Q_{SS} values are specified in the input file, and set to be a constant along the oxide-silicon interface. Fig. 9.6 shows the bird's-eye-view of the impurity

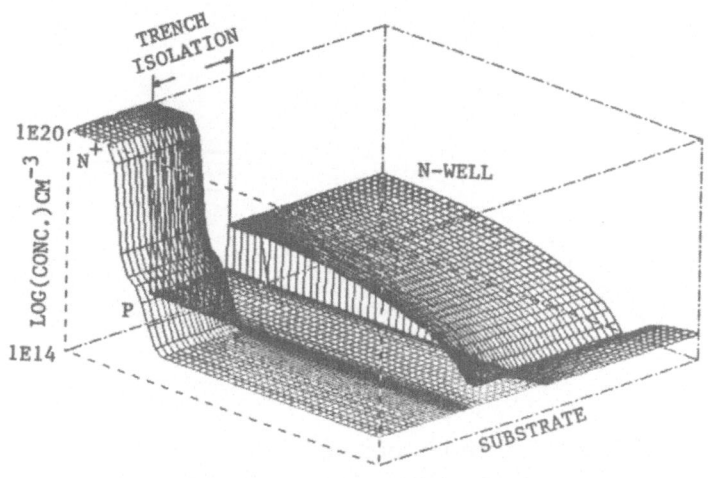

Fig. 9.6 Bird's-eye-view of the impurity profile.

profile for the trench structure. The n+ region, n-well, and p-substrate are shown. Also shown is the p-type channel implant which overlaps the n+ profile. This channel implant is used to adjust the threshold voltage of the n-channel transistor. Because of the higher p-type doping concentration near the n+ region, it will be shown that this reduces the probability of inversion near that region. The impurity concentration within the trench should be neglected in the interpretation of this plot. (The PLPKG program sets the impurity concentration within the trench equal to the substrate concentration at the bottom of the trench.)

9.3 Analysis of the Inversion Problem

In this section, the interpretation of the simulated potential distribution will be discussed. Fig. 9.7 shows a typical two-dimensional plot of the trench simulation. The constant potential contours are shown, as well as the depletion regions and p/n junctions. The potential is equal to the conduction band potential in the neutral substrate, and is set equal to the value of the applied substrate bias voltage in the neutral substrate

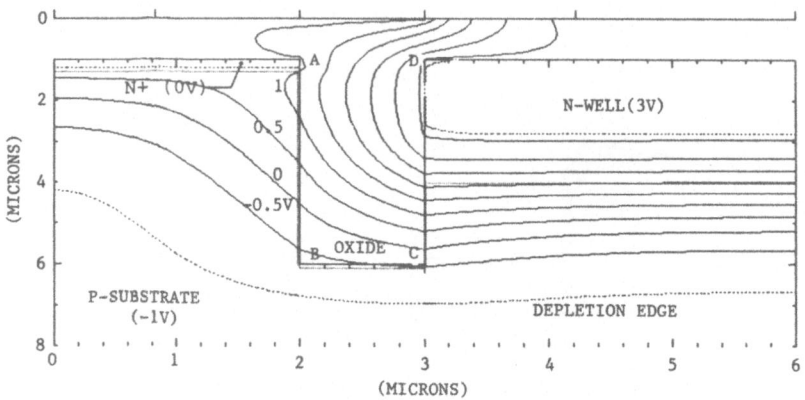

Fig. 9.7 2-D potential profile of the trench isolated
CMOS structure. The n-well bias is 3 V.

region, i.e., outside the depletion region. The bias conditions are 0, 3, -1 volts at the n+, n-well and substrate respectively. The p-type channel implant profile which overlaps the n+ region has been neglected in this case for simplicity. The shape of the potential contours indicates that the n-well and Q_{SS} both have the effect of raising the potential at the trench surface opposite to the n-well. It is interesting to look at the potential along the trench surface from the n+ to the n-well, as follows:

The path along the trench surface from the n+ to the n-well can be viewed as the channel of a MOSFET with the n+ as the source, the n-well as the gate and the drain, although part of the channel is not under the gate. From the results of the two-dimensional simulations, the potential along this path can be plotted. Fig. 9.8 shows the potential profile for different cases. The first case is for no surface charge and zero n-well bias, as a reference. The second case shows the effect of increasing the n-well bias to 3 V. The potential in the region opposite to the n-well is raised, as expected. The approximate quasi-Fermi level is also shown. The quasi-Fermi level is a function of the biases at the n+ and n-well. The condition for inversion is similar to the gated diode structure [9.10]. If the potential is above the quasi-Fermi level, then

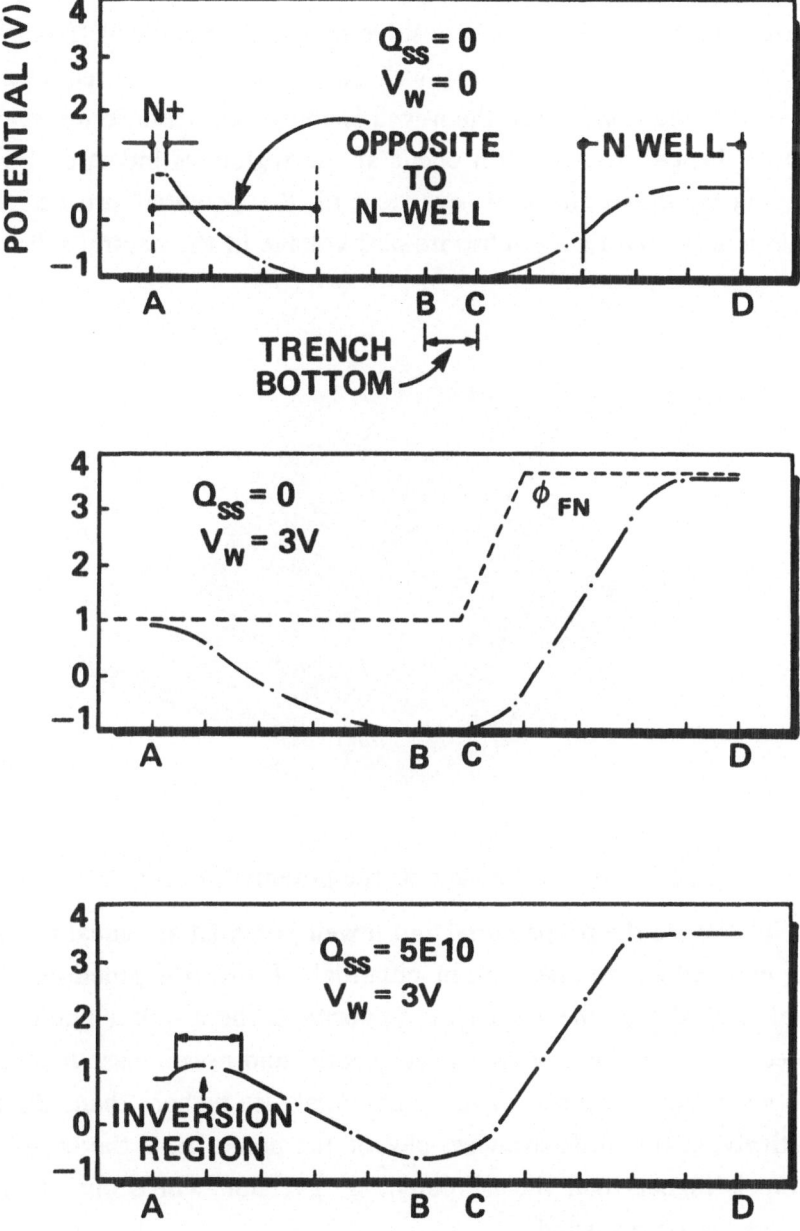

Fig. 9.8 Simulated trench surface potential from the n$^+$ region to the n-well, for different Q_{SS} and n-well bias.

inversion occurs. In this case, it is clear that no inversion has occurred, as expected. The third case is for a positive charge density of 5E10 cm^{-2} at the trench surface, and with an n-well bias of 3 V. The potential near the n$^+$ region and opposite to the n-well is above the quasi-Fermi level at the source, hence inversion will occur at the region, as indicated in the figure. Fig. 9.9 shows the bird's-eye-view of the potential profile. The potential is set equal to the substrate bias voltage in the neutral substrate

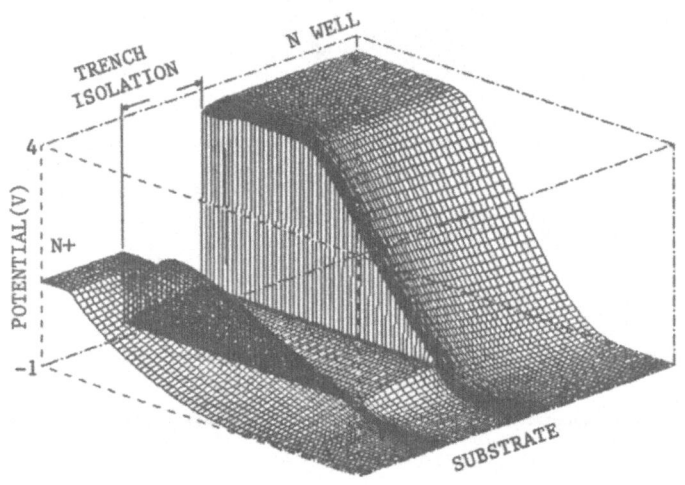

Fig. 9.9 Bird's-eye-view of the potential profile.

region. Therefore the n$^+$ potential and n-well potential are equal to their applied bias added to the built-in potential of the p/n junction. The potential "bump" near the source and opposite to the n-well is due to the combined effect of the positive surface charge and n-well bias. Inversion occurs when the trench surface potential is higher than $2\psi_B$. Qualitatively, in the bird's-eye-view plot of the potential, if the height of the bump is higher than the n$^+$ potential, inversion will occur. In this case, inversion has occurred.

The effect of the p-type channel implant on the inversion problem is also studied. Fig. 9.10 shows the vertical impurity profile at the n$^+$

region. The p-type channel implant has a significant effect on the trench surface inversion near the n+ region, since it increases the p-type impurity concentration near that region. Fig. 9.11 shows the GEMINI simulation

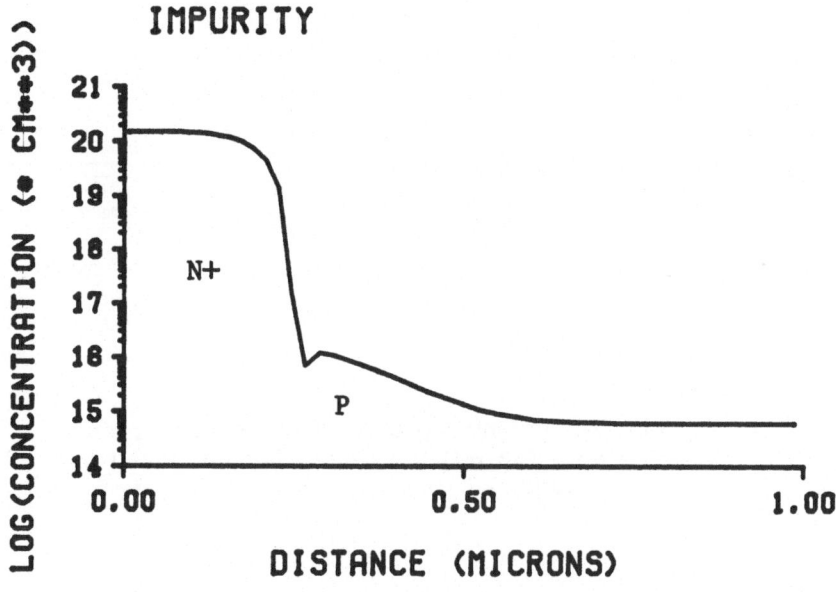

Fig. 9.10 Vertical impurity profile at the n+ region.

where the p-type channel implant which overlaps the n+ region is taken into account. The two-dimensional profile shows a reduction of the potential in that region. In this case, the probability of inversion is reduced, under the same conditions. Fig. 9.12 shows the bird's-eye-view of the potential profile. The magnitude of the potential "bump" is reduced, in comparison to Fig. 9.9. The potential is also suppressed at the edge of the n+ region due to the higher boron concentration. Note that this channel implant will be increased in scaled down devices to prevent drain-induced barrier lowering problems, as well as to adjust the threshold voltage for thinner gate oxide. Therefore, the inversion problem near the n+ region will be reduced for scaled CMOS. The trench surface inversion in the bulk, however, will be dependent only on the charge density and the substrate doping concentration.

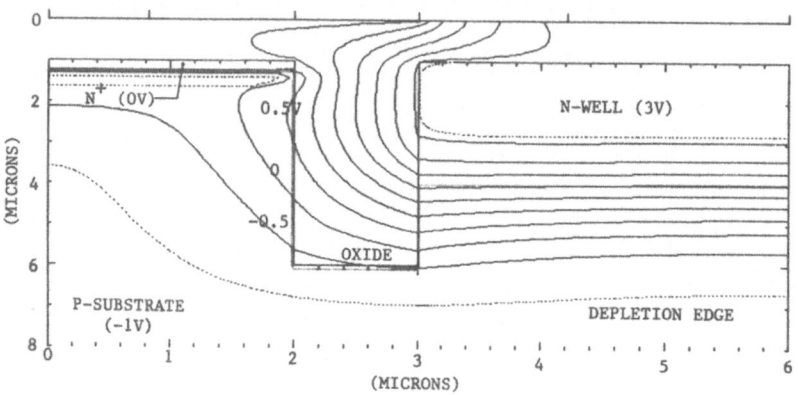

Fig. 9.11 2-D potential profile of the trench structure, with the p-type channel implant taken into account.

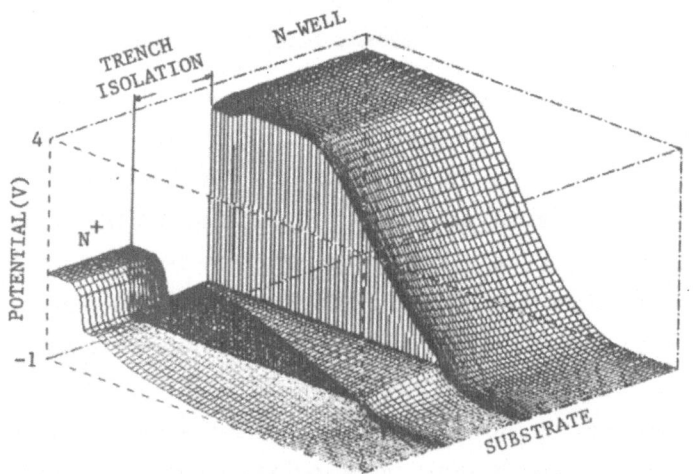

Fig. 9.12 Bird's-eye-view of the potential profile. The p-type channel implant is taken into account.

9.4 Summary Of Simulation Results

The trench simulation has been performed for different values of Q_{SS} and bulk doping concentration, n-well bias, and substrate bias. The value of Q_{SS} ranges from 0 to 3E11 cm^{-2}, and the bulk doping concentration ranges from 6E14 to 1E16 cm^{-3}. N-well biases of 0 V and 3 V and substrate bias of 0 to -3 V are considered. Table 9.1 summarizes the

QSS (/CM2)	NA (/CM3)	VW (V)	VB (V)	INVERSION
0	6E14	3	−1	NO
5E10	6E14	3	−1	NO
1E11	6E14	3	−1	YES
1E11	2E15	3	−1	NO
3E11	5E15	3	−1	YES
3E11	1E16	3	−1	NO
1E11	6E14	3	0	YES
1E11	6E14	3	−1	YES
1E11	6E14	3	−3	YES
1E11	6E14	0	−1	NO

QSS=POSITIVE FIXED CHARGE DENSITY
NA=BULK DOPING CONCENTRATION
VW=N−WELL BIAS; VB=SUBSTRATE BIAS

Table 9.1 Summary of trench simulations.

results of the simulations. The cases to be considered are put into four groups. The first group studies the effect of increasing Q_{SS}. The second group studies the effect of increasing substrate doping. The third group studies the effect of substrate bias and the last studies the effect of n-well bias. Also indicated in the last column is a qualitative description of the simulated results. The n^+ region is biased at 0 V. In all cases, the p-type channel implant overlapping the n^+ region is taken into account.

From the first three rows of Table 9.1, the results show that with a substrate bias of -1 V, and p-substrate doping of 6E14 cm^{-3}, trench surface inversion occurs if Q_{SS} is greater than 5E10 cm^{-2}. The second group, from the fourth to sixth row, shows that for a Q_{SS} of 1E11 cm^{-2} and 3E11 cm^{-2}, the substrate doping has to be raised to 2E15 cm^{-3} and 1E16 cm^{-3} respectively in order to avoid trench surface inversion. The substrate bias effect is shown from the 7th to the 9th row. The substrate bias shows little effect in preventing inversion if Q_{SS} is 1E11 cm^{-2} and p-substrate doping is 6E14 cm^{-3}, although it is found that at lower Q_{SS} values, negative substrate bias reduces the probability of inversion. The last row shows that for a Q_{SS} value of 1E11 cm^{-2}, the inversion can be avoided if n-well is biased at zero volt. This merely serves to reveal the effect of the n-well bias. The n-well has to be biased to a positive voltage for the circuit to function, and the results show that minimizing the n-well bias will reduce the probability of inversion.

9.5 Experimental Results

The trench surface inversion problem has been studied experimentally and compared with the simulated results. The problem of trench surface inversion is manifested in the subthreshold characteristics of the n-channel transistor, as shown in Fig. 9.13. The transistor which is essentially adjacent to the trench shows severe leakage problem, while the transistor at a distance of 1.5 μm away from the trench has no leakage problem, and is electrically indistinguishable from LOCOS isolated transistors. The p-channel transistor has no leakage problem, and is similar to non-trench devices. This indicates that the problem is positive charge on the trench surface or in the refilled material which causes inversion along the trench surface in the p-substrate region.

The effective charge density at the trench surface was determined by the trench surface inversion test structures [9.7]. The value was found to

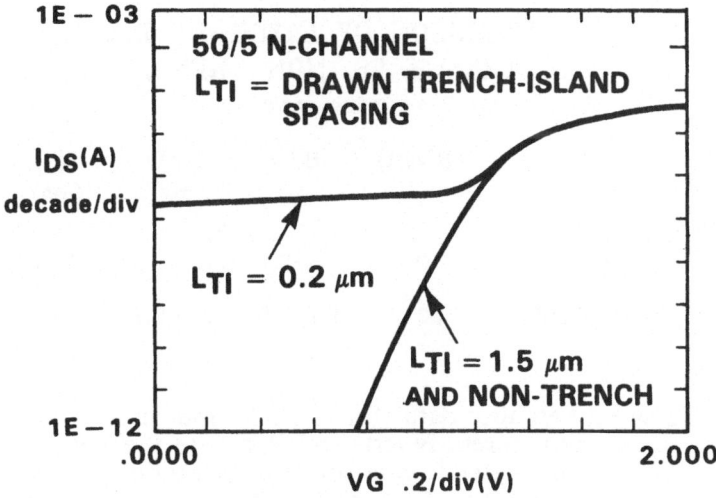

Fig. 9.13 Subthreshold behavior of trench isolated
n-channel MOSFET, with different trench to
island spacing. Non-trench device data is also
shown for comparison.

be 1.8E11 cm^{-2}. Comparison with Table 9.1 of the simulation results
shows that a p-substrate doping concentration of at least 5E15 to
1E16 cm^{-3} is necessary to avoid inversion, which means that p-well
CMOS is a better choice for the application of trench isolation.

Experimental results have also shown that when the trench surface is
inverted, the latchup resistance of the trench structure is not as high as
expected. Table 9.2 shows the results for the measurements of the
latchup initiating and holding current levels for the trench and
non-trench structures. The difference is quite small, although the trench
structure has an n$^+$ to p$^+$ spacing of 4 μm, which is much smaller than
the conventional test structure, of 8 μm. The latchup resistance of the
trench structure is much less than expected in this case, and is due to the
existence of the inverted channel along the trench, all the way to the
n-well. If this conducting path to the n-well is disconnected, for example,
by a p-type implant into the trench [9.4], the latchup resistance will be

COMPARISON OF LATCH–UP DATA
TRENCH VS. NON–TRENCH
(CT029–12,17)

	B(vert)	B(lat)	I(in) (mA)	I(h) (mA)
TRENCH	30	6–9	1.5–2.5	3–4
NON–TRENCH	50	4.5–7.5	0.3–0.5	2–2.5

Table 9.2 Latchup data for trench and non-trench structures. B(vert) and B(lat) are the vertical and lateral gain of the parasitic bipolar transistors. I(in) and I(h) are the latchup initiating and holding currents.

recovered and would be much better than the conventional LOCOS structures.

9.6 Summary

By using two-dimensional simulations, the problem of the trench surface inversion has been studied in detail to investigate the possibility of placing the transistors adjacent to the trench. Different conditions of charge density, biases, and doping concentrations have been considered. The regions of inversion have been identified. The results show that if Q_{SS} is larger than 5E10 cm^{-2}, then inversion will likely occur, unless the p-substrate doping concentration is increased. If Q_{SS} is larger than 1E11 cm^{-2}, then p-well CMOS is more suitable for the application of trench isolation, due to the much higher doping concentration in the p-well. By minimizing the n-well bias, the probability of inversion is also reduced. Experimental results indicated Q_{SS} values of about 2E11 cm^{-2}. Latchup resistance of the trench isolated structures with the transistors adjacent to the trench side-wall showed only slight improvement over the

LOCOS isolated structures, under the condition of surface inversion. This indicates that the inversion path between the n^+ region and the n-well has to be disconnected in order to recover the trench resistance to latchup. This can be done either by implanting into the trench, or by using p/p^+ epi where the trench bottom is within the p^+ region. Trench isolation is more suitable for p-well CMOS, where the inversion problem can be put under control if Q_{SS} is not too large. If body effect and diffusion capacitance are not of major concern, then trench isolation can also be used for n-well CMOS or NMOS processes by using a lower resistivity substrate, or using high dose and deep boron channel implants. If inversion is a major problem, then the n-channel transistor has to be put at a distance of about 1 μm away from the trench, where experimental results showed that leakage is not a problem.

References:

[9.1] S.-Y. Chiang, K. M. Cham, D. W. Wenocur, A. Hui, and R. D. Rung, "Trench Isolation Technology for MOS Applications," *Proc. of the First International Symposium on VLSI Science and Technology,* 1982, pp. 339-346.

[9.2] R. D. Rung, H. Momose, and Y. Nagakubo, "Deep Trench Isolated CMOS Devices," *Tech. Digest of IEDM 1982,* pp. 237-240.

[9.3] K. M. Cham and S.-Y. Chiang, "A Study of the Trench Surface Inversion Problem for the Trench CMOS Technology," *IEEE Electron Device Lett.,* EDL-4, Sept 1983, pp. 303-305.

[9.4] T. Yamaguchi, S. Morimoto, G. H. Kawamoto, H. K. Park, and G. C. Eiden, "High Speed Lathup-Free, 0.5 μm-Channel CMOS Using Self-Aligned TiSi$_2$ and Deep Trench Isolation," *Tech. Digest of IEDM 1983,* pp. 522-525.

[9.5] D. B. Estreich, "The Physics and Modeling of Latch-Up in CMOS Integrated Circuits," Stanford Electronics Labs Report #G-201-9, 1980.

[9.6] R. R. Troutman, "Recent Developments in CMOS Latchup," *Tech.*

Digest of IEDM 1984, pp. 296-299.

[9.7] K. M. Cham, S.-Y. Chiang, D. W. Wenocur, and R. D. Rung, "Characterization and Modeling of the Trench Surface Inversion Problem for the Trench Isolated CMOS Technology," *Tech. Digest of IEDM 1983*, pp. 23-26

[9.8] C. W. Teng, C. Slawinski and W. R. Hunter, "Defect Generation in Trench Isolation," *Tech. Digest of IEDM 1984*, pp. 586-589.

[9.9] S. M. Sze, *Physics of Semiconductor Devices*, second ed., New York:Wiley-Interscience, 1981.

[9.10] A. S. Grove, *Physics and Technology of Semiconductor Devices*, New York:Wiley, 1967.

Chapter 10

Development of Isolation Structures for Applications in VLSI

10.1 Introduction to Isolation Structures

Device isolation has become one of the major issues in VLSI. As more and more devices are packed together on a single chip, the spacing between devices is reduced significantly. Island width/space design rules are becoming very aggressive, in the range of 1 μm/1 μm. [10.1] This means that the width of the isolation structures has to be scalable without causing field leakage problems. Also, as the transistor widths are scaled down, to the range of 2 μm or less, narrow width effects become a major issue. [10.2]-[10.5]. These effects are dependent on the isolation structures since the channel width of the device is defined by the field isolation. Many novel isolation structures have been investigated for applications in VLSI [10.6]-[10.14].

In general, there are several considerations in the design of isolation structures. First, the isolation dielectric thickness should be as thick as possible. This will provide a higher field threshold voltage, as well as a lower parasitic capacitance for conducting lines over the field oxide. Second, the field isolation should not encroach significantly into the island area (active area). In VLSI, over a million transistors may be packed on a chip, which means that the transistor size is a major consideration. Drawn channel widths (i.e., mask dimensions) are being

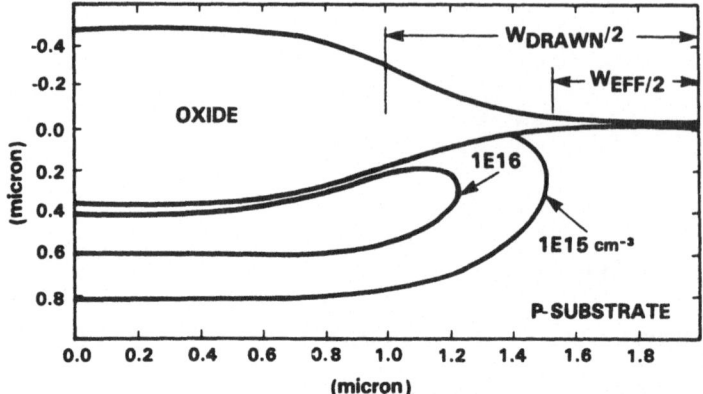

Fig. 10.1 Field oxide and boron encroachment in LOCOS.

scaled   to   the   1 μm   range   in   advanced   integrated   circuit development [10.15]. Any encroachment of the field isolation into the active area will cause a significant reduction in the current drive of the device. This problem is shown schematically in Fig. 10.1 for the case of LOCOS isolation. [10.16].   The effective channel width of the device is significantly smaller than the drawn width. The slope of the field oxide at the boundary of the active area is also a consideration. If the slope is shallow as is the case for LOCOS, then this causes an increase in the threshold voltage of the transistor as the device width is scaled down. The threshold increase occurs because the gate electric field spreads into the field area, causing an increase of the effective channel doping of the device. Also, if a lightly doped substrate is used, a field implant of an impurity of the same type as the substrate is necessary to produce the desired field threshold voltage. In this case, the impurity may diffuse into the active area during processing, thus causing an increase in the threshold voltage or a loss of effective channel width of the active transistor. These problems of oxide and boron encroachments will be discussed in more detail later for the case of LOCOS isolation. Third, the isolation structure is preferred to be planar with respect to the silicon surface so that subsequent steps in photolithography and etching will not be affected by topological problems. A planar surface, for example, will minimize necking of narrow polysilicon lines running across an island/field edge. The last consideration, but not the least, is process complexity. It is highly desirable that any novel isolation structures be compatible with existing MOS processes, and not require additional masking steps.

From the above discussions, it can be seen that in most cases, compromise must be made to optimize the design of the isolation structure. Simple isolation techniques such as LOCOS result in oxide encroachment. Higher implant dose to maintain the field threshold results in higher diffusion capacitance. The use of thicker field oxide to reduce interconnect capacitance may result in more oxide encroachment as well as more complicated processing problems due to topology. The following sections describe the design of isolation structures, using simulation programs as tools to optimize the design. The discussion begins with the simplest isolation technique, LOCOS. Then, modifications of this technique to minimize oxide and boron encroachment are discussed. Finally, a new isolation technique, SWAMI, is presented, which will be shown to solve most of the concerns mentioned above.

10.2 Local Oxidation of Silicon (LOCOS)

LOCOS [10.16] has been used in most integrated circuits up till this date. The process is very simple and well known. Only a brief description will be presented. The process is shown in Fig. 10.2, using the output of the SUPRA simulation. A layer of thin oxide (stress relief oxide) is grown above the silicon substrate. This is followed by a layer of nitride. Then the nitride is patterned using photolithography technique, followed by plasma etching. This defines the regions of field and active area. Then a field implant is performed to increase the threshold voltage at the field. The photoresist is then removed followed by oxidation of the exposed silicon area. The simulation shows that the encroachment of the oxide is very significant. For this case, the encroachment is about 0.3 μm on each

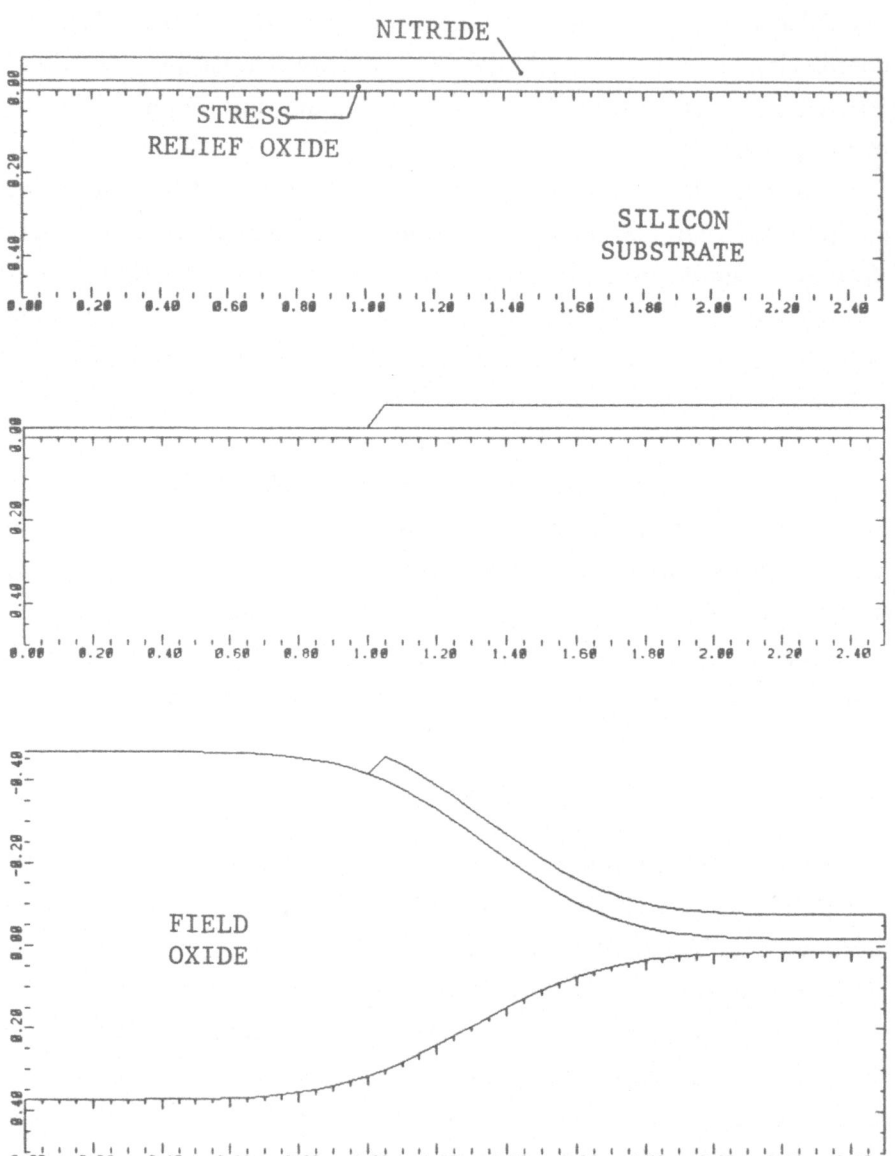

Fig. 10.2 LOCOS process simulated by SUPRA.

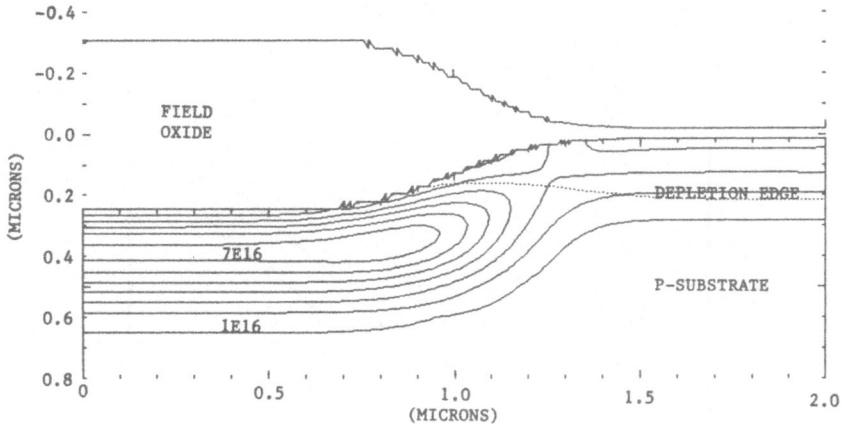

Fig. 10.3 LOCOS isolated channel width simulation. The boron distribution due to the field and channel implants, as well as the depletion edge are shown.

side for a field oxide thickness of 0.55 μm. This oxide encroachment is mainly due to the presence of the stress relief oxide underneath the nitride, which is necessary to avoid defects in the active area of the silicon. The oxygen diffuses through this oxide layer during field oxidation, and hence oxide is grown under the nitride layer near the field edge.

Another concern with LOCOS is the encroachment of the boron impurities into the active area. If the impurity concentration underneath the active area is significantly changed, then this will cause a reduction of effective channel width, and may also cause an increase in the threshold voltage of the active transistor. In order to see how serious is the problem, the SUPRA program is used to simulate the 2-D structure, and the output data is coupled to the GEMINI program to calculate the current-voltage characteristics of the transistor in the linear region. Fig. 10.3 shows the simulated cross-section of the width of the transistor. The LOCOS isolation structure together with the field and channel

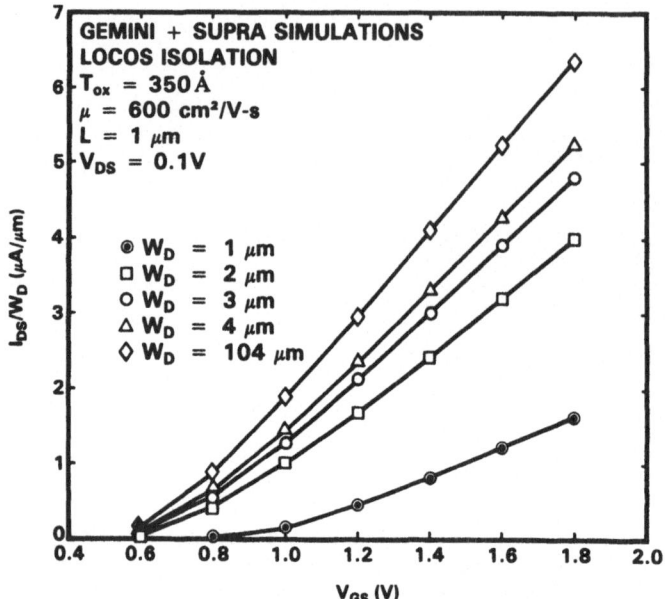

Fig. 10.4 I_{DS}/W_D vs. V_{GS} for LOCOS isolation.

implant impurity profile are shown.   The simulated depletion region
indicates that part of the gate electric field has terminated into the field
region, hence there will be an effect on the electrical characteristics of
the device.

Fig. 10.4   shows   the   simulated   current-voltage   characteristics   of
transistors with varying drawn widths (W_D). I_{DS}/W_D is plotted versus
V_{GS} for drawn channel widths from 1 to 4 µm as well as for a very
wide device of 104 µm. If the isolation is ideal, such that the electrical
channel width and the drawn channel width are the same, then all of the
curves   will be   identical   to   that   of   the   wide   device,   since I_{DS} is
proportional to the effective channel width. The simulated results show
that for LOCOS isolation, this is not the case.   The value I_{DS}/W_D
decreases rapidly as W_D is reduced to 1 µm. This clearly shows that
LOCOS is not acceptable for VLSI.

The effective width of the transistors can be calculated from these simulated results. Since the transconductance, given by the slope of the I_{DS} vs. V_{GS} curve, is proportional to the effective channel width of the device, a plot of the transconductance versus the drawn channel width will provide us the value of the channel width loss due to the isolation process. Fig. 10.5 shows the extraction of the channel width. The actual width is found to be 0.7 μm less than the mask width.

The channel width loss can be due to both the oxide encroachment and the boron field implant which diffuses into the active area, causing an increase of the threshold voltage at the active area edge. The latter effect can be studied by comparing the simulated current-voltage characteristics of a narrow width device with and without field implant. Fig. 10.6 shows that the field implant only reduces the current slightly, hence the device channel width loss is mainly due to the oxide encroachment. In the case where oxide etch back is used to reduce the bird's beak, the effect of boron encroachment will be more significant. This will be discussed in the next section.

Another figure of merit for an isolation technology is the sensitivity of the threshold voltage to the device width. From the circuit design point of view, it is highly desirable that the threshold voltage be insensitive to the device geometry. Fig. 10.7 shows the simulated results of the threshold voltage versus the drawn channel width for the LOCOS isolation. The threshold voltage begins to increase rapidly for drawn width below 2 μm. This is because the gate electric field partially terminates under the field oxide at the edge of the active area. This can be observed in Fig. 10.3 which shows the channel depletion edge. This again indicates that the application of LOCOS is limited to a minimum device width of at least 1.5 μm for optimum performance.

10.3 Modified LOCOS

One way to reduce the oxide encroachment (bird's beak) is to grow a thick field oxide and then etch back to a thinner thickness. The idea can be understood by looking at a field oxide profile generated by a two

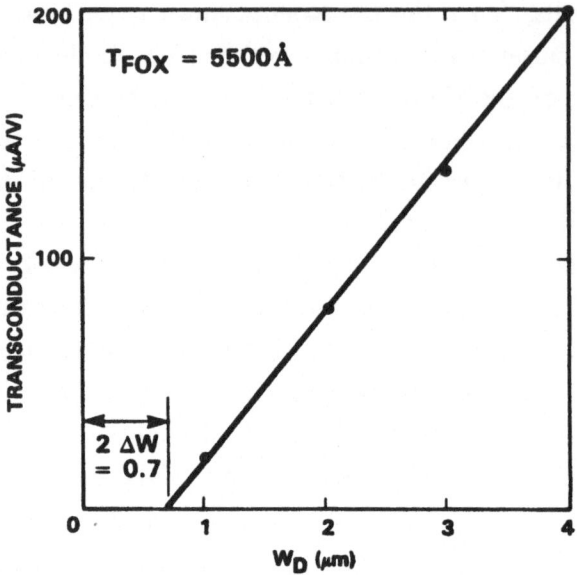

Fig. 10.5 Extraction of the effective channel width (W_D-2ΔW).

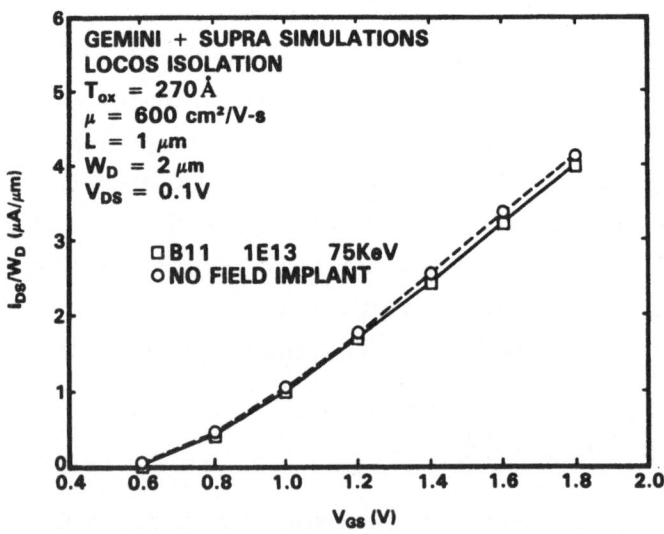

Fig. 10.6 Simulation of the I_{DS}-V_{GS} characteristics for LOCOS isolated MOSFET, with and without field implant.

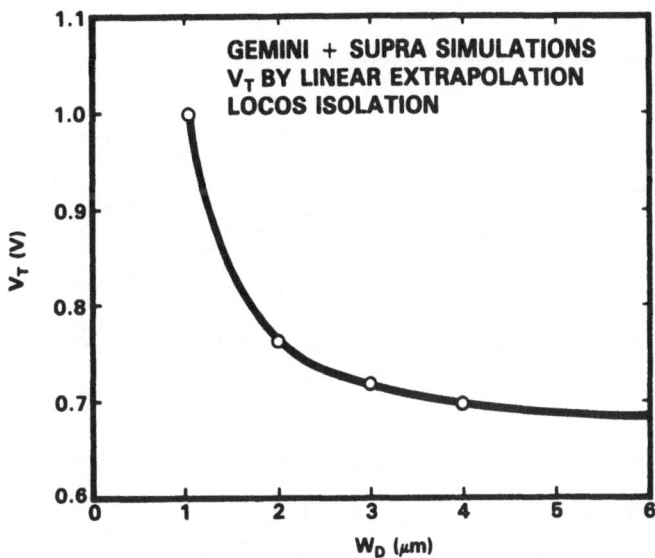

Fig. 10.7 Simulated threshold voltage vs. W_D.

dimensional oxidation program SUPRA, as shown in Fig. 10.2. It can be seen from the profile that the bird's beak is long but the initial slope is quite flat. Etching away 200 nm to 300 nm of field oxide can hence reduce the bird's beak significantly. In this section, an optimum design of this etch back LOCOS process using one-dimensional process simulation program SUPREM and two-dimensional simulation program SUPRA will be discussed. The goal is to minimize both the boron and oxide encroachments.

The original process uses a stress-relief oxide thickness of 40 nm, a nitride thickness of 200 nm and a boron field implant dose of $5E13$ cm^{-2} to obtain a final field threshold voltage of about 15 V. 800 nm of field oxide is grown at 950 deg C in steam for 300 minutes and then etched back to 500 nm. Fig. 10.8 shows the simulated final field oxide and doping profile from SUPRA. The vertical line indicates the position of the bird's beak before etch back. The oxide encroachment is reduced by about 0.28 microns and the active area is about the same as originally defined by the nitride. However, the boron encroachment effect is quite

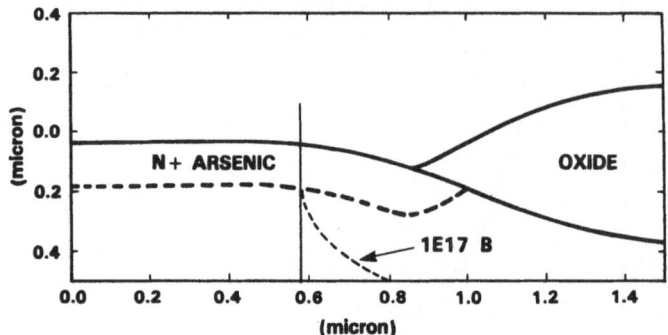

Fig. 10.8 Oxide and impurity profile after etch back.

significant. The dotted line in Fig. 10.8 shows that due to field implant the boron concentration line of 1E17 cm^{-3} is quite a distance into the active area. Simulation of the active area boron concentration due to threshold and punchthrough control implant is plotted as a function of distance into the silicon substrate in Fig. 10.9. The surface boron

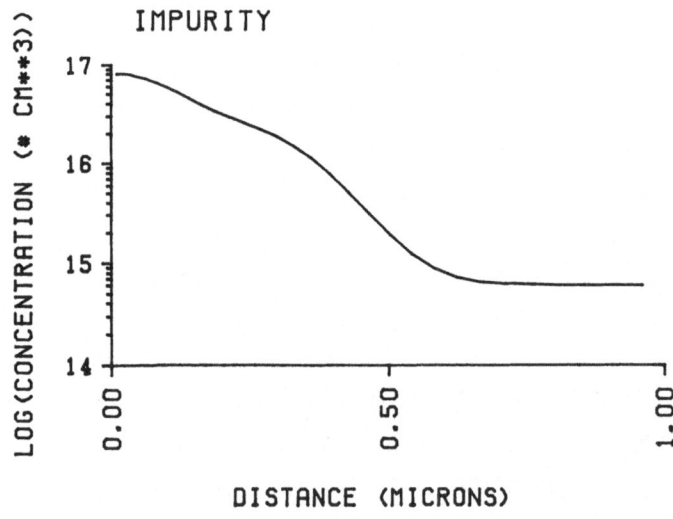

Fig 10.9 Channel impurity profile vs. vertical distance.

concentration is about 8E16. Therefore, a significant part of the active area has a higher boron concentration due to the lateral diffusion of the boron field implant. In this case, the boron encroachment actually dominates over the oxide encroachment. The effective channel width is not much improved by this etch back process.

In order to solve the boron encroachment problem the process is modified and simulated again. A thinner field oxide is grown so that both the thermal cycle and the field implant boron dose can be reduced. In the modified process, 600 nm of field oxide is grown and then etched back to about 500 nm. SUPREM is used to find the oxidation cycle for 600 nm of field oxide and the boron dose needed to give the same field threshold voltage of 15 V. It is found that the field implant dose can be reduced from 5E13 to 3E13 cm^{-2} and the thermal cycle can be reduced to 200 minutes at 950 deg C. SUPRA is then used to simulate the entire new process. The resulting field oxide and doping profile is shown in Fig. 10.10. The field oxide encroachment in this case is about 0.15 micron more than in the previous process. However, the boron encroachment is

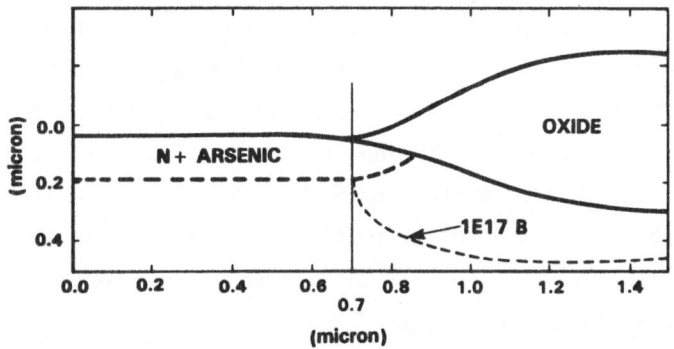

Fig. 10.10 Oxide and impurity profile for modified LOCOS.

about 0.13 micron less, using the 1E17 cm^{-3} boron line as a reference. Overall, there is a gain of 0.1 micron over the previous process. Hence, in the design of a optimum LOCOS process, both the oxide and boron encroachment need to be carefully taken into account.

10.4 Side Wall Masked Isolation (SWAMI)

The problems of oxide and boron encroachments associated with LOCOS must be minimized in VLSI. New isolation techniques have to be designed such that the considerations discussed in the introduction are satisfied in order to achieve optimum circuit performance. A new isolation process, SWAMI (Side Wall Masked Isolation), has been developed at Hewlett-Packard Laboratories [10.8]. This process effectively eliminates the bird's beak during field oxidation while maintaining a desirable field oxide thickness.

Fig. 10.11 shows the sequence of steps in the fabrication process used to form the SWAMI isolation structure. For more detailed descriptions, the reader is referred to the reference article. The objectives of the process are to prevent field oxide encroachment into the active area by using a second layer of thin nitride on the side wall and to reduce stress by using a sloped side wall. The sloped side wall is used to prevent defect formation due to volume expansion induced stress during field oxidation. SEM cross-sectional view of the isolation structure did not reveal any bird's beak formation in the active area. The sloped side wall is produced by C_2F_6 plasma etching [10.8], which introduces a loss of island width of about 0.1 μm per side. Another factor that might contribute to the channel width loss is the boron encroachment, since a field implant step is necessary to eliminate the problem of field inversion, especially at the sloped side wall. In this section, simulations will be used to compare the performance of narrow width devices using an ideal vertical isolation, the SWAMI, and LOCOS. The effect of boron encroachment for the SWAMI process is also simulated.

To simulate the SWAMI oxide structure, a program with the capability of simulating generalized isolation structures is needed. This can be done by the combination of SUPRA and SOAP. The SOAP program is used to generate the isolation structure. The output data file is then loaded into the SUPRA file for the impurity profile simulations. Fig. 10.12 shows the isolation structure generated by SOAP for the SWAMI process.

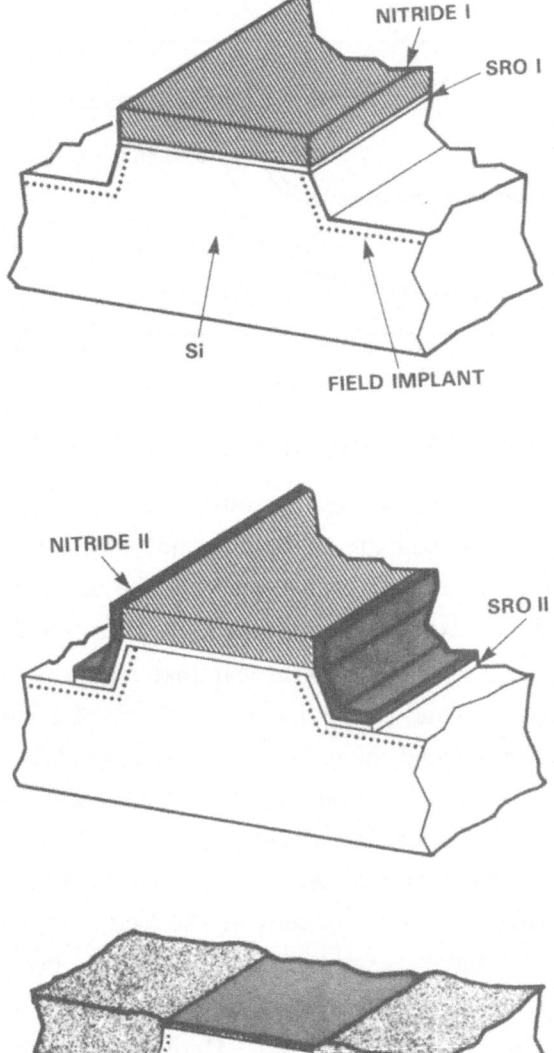

Fig. 10.11 The SWAMI process.

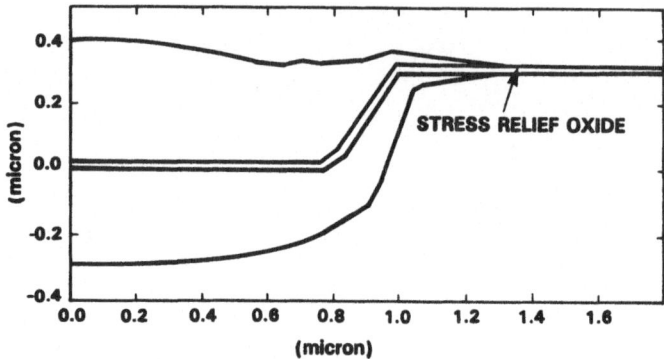

Fig. 10.12 SWAMI structure simulated by SOAP.

The sloped silicon side wall and the nitride layer covering the side wall
have significantly reduced the oxide encroachment into the active area,
although there is still a bird's beak formed at the island edge. Since SEM
pictures have shown experimentally that there is essentially no bird's beak
at the island edge, the oxide growth near that region is overestimated by
the SOAP program. One possibility is the effect of stress in that region
which might have reduced the oxygen diffusivity [10.17].

The SOAP program also simulates the stress along the oxide/silicon
interface as a function of time during oxidation. This is useful if one is
concerned about the stress induced by the isolation structure. In the case
of SWAMI, stress may be a concern at the island edge covered by the
nitride, where volume expansion induced stress occurs during field
oxidation. Fig. 10.13 shows the stress calculations using the SOAP
program for the SWAMI process. Here the positive stress means
compression and the negative stress dilation. The stress calculated is after
0.5 μm of field oxide is grown. The horizontal axis corresponds to the
horizontal distance from the left edge of the structure. The results show
that stress occurs along the sloped side wall covered by the thin nitride.
As oxide is grown, stress is induced by volume expansion. The results also

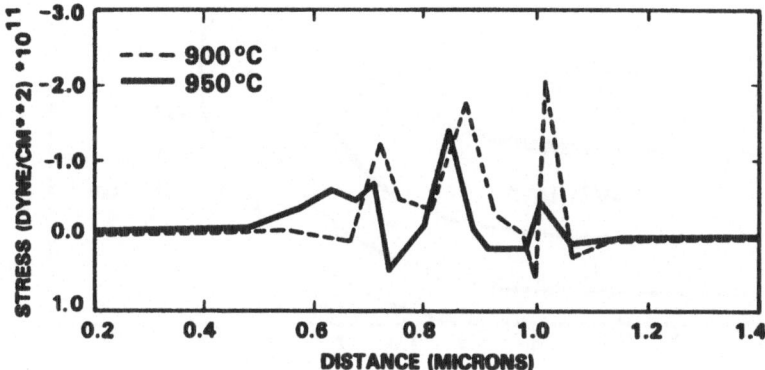

Fig. 10.13 Stress simulation for SWAMI using SOAP.

show that stress can be reduced by using higher oxide growth temperature. This is due to the lower viscosity of the oxide, hence more flow is possible to reduce the stress.

In order to see if boron encroachment is a problem in SWAMI, the impurity distribution has to be simulated. This can be done using the SUPRA program. The program accepts the data file from the SOAP program which defines the final oxide structure. The SUPRA program then calculates the impurity distribution associated with the oxide boundary conditions and the temperature cycle of the field oxidation. Then the channel implants and subsequent process steps can follow. Fig. 10.14 shows the simulated structure with the impurity contours without the channel implants. This shows the extent of the boron distribution due to field implant. From this figure, it can be seen that the boron concentration under the active area is not high, in comparison with typical channel implants, especially for submicron devices (Fig. 8.17). The effect on the threshold voltage and channel width narrowing should be only slight.

Since the SOAP program overestimates the oxide encroachment, the device performance simulated by using this structure will not be accurate. The electrical effects of the oxide shape can be approximated by the

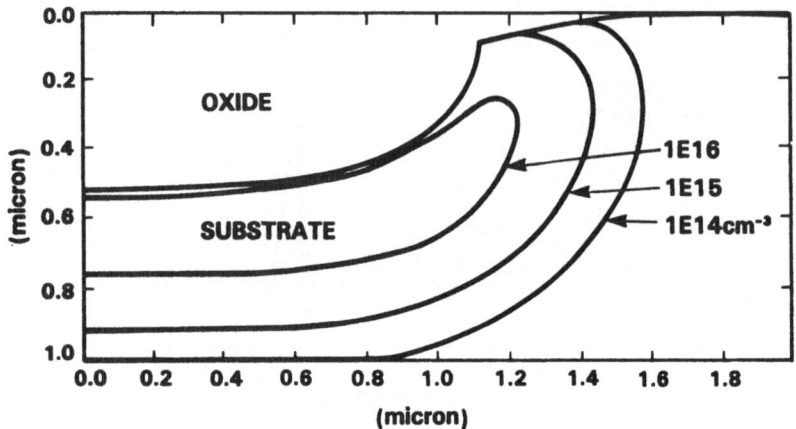

Fig. 10.14 Impurity profile simulation for SWAMI.

GEMINI program. If the depletion region of the device does not extend
too deeply into the substrate, then it is only necessary to approximate the
side wall of the oxide structure. The impurity profile of the structure
can be first simulated by SOAP-SUPRA combination, then copied into
the GEMINI input file by using a Gaussian profile approximation. This
procedure is somewhat lengthy, but should provide better agreement with
experiments.

Fig. 10.15 shows the simulation of the SWAMI structure for a
MOSFET with channel width of 2 μm. Only half of the device needs to
be simulated due to symmetry. The GEMINI simulation provides a good
representation of the upper part of the oxide structure, which determines
most of the narrow width effects. Note that the depletion region ends
near the top of the isolation, hence any approximation of the oxide shape
near the bottom will not give rise to significant errors. The drain to
source current as a function of the gate bias can be simulated. From the
current-voltage characteristics, the threshold voltage as well as the
transconductance can be calculated. This provides information about the
narrow width effects, which are manifested as increase in threshold
voltage and reduction in transconductance.

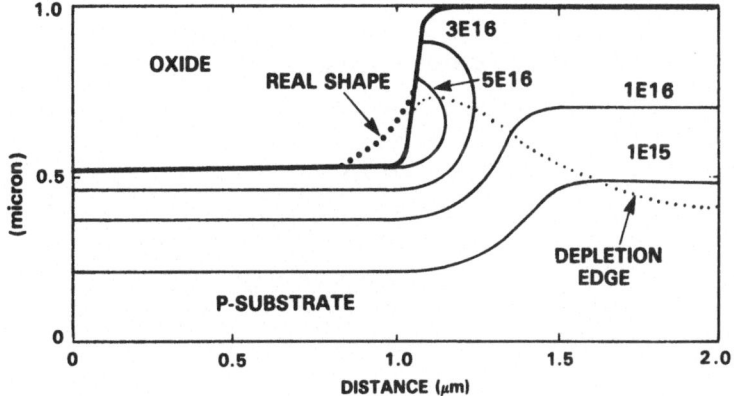

Fig. 10.15   Channel width simulation of SWAMI by GEMINI.

This procedure is repeated for the cases of an ideal vertical isolation, and also for LOCOS. The effect of boron encroachment can be studied by simulating structures with and without field implants. Oxide thickness of 500 nm is considered. Fig. 10.16 shows the simulated current voltage characteristics of the different cases. I_{DS} is plotted versus V_{GS} for the devices biased in the linear region. From the slopes of the curves, the effective width of the devices can be calculated. SWAMI is slightly inferior to the ideal vertical isolation, but much better than LOCOS. Note that the main effect of the boron field implant on the SWAMI structure is to increase the threshold voltage, while the transconductance is not significantly changed. For the case of LOCOS, the reduction in current is mostly due to the oxide encroachment into the active area. The effect of boron encroachment is insignificant. For the SWAMI process, the channel width loss is almost independent of the field oxide thickness up to about 0.7 μm, while the LOCOS has a width loss per side approximately 70% of the oxide thickness. Hence for thicker field oxide, which is desirable for reducing the interconnection parasitic capacitance, isolation structures such as SWAMI are absolutely necessary. Table 10.1 summarizes the results of the simulations. The channel width loss for the

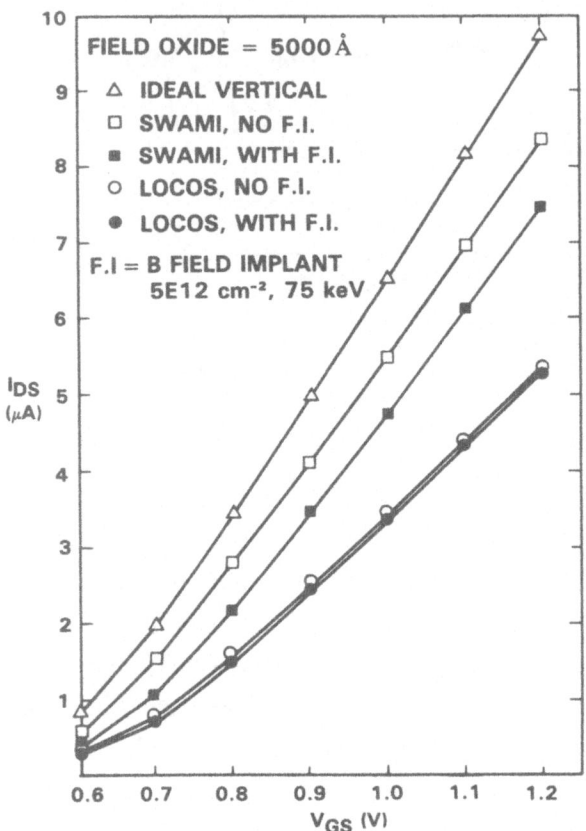

Fig. 10.16 Simulated I_{DS} - V_{GS} characteristics of n-channel MOSFET with drawn width of 2 μm, for vertical, SWAMI and LOCOS isolations.

SWAMI process is found to be 0.11 μm per side due to oxide encroachment, and about 0.02 μm per side due to boron encroachment.

10.5 Summary

Simulations have been found to be very useful in developing and understanding isolation structures. The electrical characteristics of narrow width devices can be understood and optimized by coupling the process simulations and device simulations. Advanced isolation structures such as SWAMI can be simulated; SWAMI's effect on device characteristics has been studied and compared with other isolation structures. As isolation structures evolve to higher complexity, the use of numerical simulations is essential for efficient process and device development.

ISOLATION STRUCTURE	DW (MICRONS) PER SIDE
IDEAL VERTICAL	0
LOCOS, NO F.I.	0.35
LOCOS, WITH F.I.	0.36
SWAMI, NO F.I.	0.11
SWAMI, WITH F.I.	0.13

F.I.= BORON FIELD IMPLANT
FIELD OXIDE THICKNESS = 5000A

Table 10.1 Summary of MOSFET width simulations.

References

[10.1] O. Kudoh, H. Ooka, I. Sakai, M. Saitoh, J. Ozaki and M. Kikuchi, "A New Full CMOS SRAM Cell Structure," *Tech. Digest of IEDM 1984*, pp. 67-70.

[10.2] T. Yamaguchi and S. Morimoto, "Analytical Model and Characterization of Small Geometry MOSFET's," *IEEE Trans. on Electron Devices*, ED-30, June 1983, pp. 559-571.

[10.3] L. A. Akers, M. M. E. Beguwala, and F. Z. Custode, "A Model of a Narrow-Width MOSFET Including Tapered Oxide and Doping Encroachment," *IEEE Trans. Electron Devices*, ED-28, Dec 1981, pp. 1490-1495.

[10.4] C.-R. Ji and C.-T. Sah, "Analysis of the Narrow Gate Effect in Submicrometer MOSFET's," *IEEE Trans. Electron Devices*, ED-30, Dec 1983, pp. 1672-1677.

[10.5] F. H. Gaensslen, "Geometry Effects of Small MOSFET Devices," *IBM J. Res. Develop.*, 23, Nov 1979, pp. 682-688.

[10.6] R. D. Rung, H. Momose, and Y. Nagakubo, "Deep Trench Isolated CMOS Devices," *Tech. Digest of IEDM 1982*, pp. 237-240.

[10.7] K. L. Wang, S. A. Saller, W. R. Hunter, P. K. Chatterjee, P. Yang, "Direct Moat Isolation for VLSI," *IEEE Trans. Electron Devices*, ED-29, Apr 1982, pp. 541-547.

[10.8] K. Y. Chiu, J. L. Moll, K. M. Cham, J. Lin, C. Lage, S. Angelos, and R. C. Tillman, "The Sloped-Wall SWAMI--A Defect-Free Zero Bird's Beak Local Oxidation Process for Scaled VLSI Technology," *IEEE Trans. Electron Devices*, ED-30, Nov 1983, pp. 1506-1511.

[10.9] N. Matsukawa, H. Nozawa, J. Matsunaga, S. Kohyama, "Selective Polysilicon Oxidation Technology for VLSI Isolation," *IEEE Trans. Electron Devices*, ED-29, Apr 1982, pp. 561-567.

[10.10] S. Hine, T. Hirao, S. Kayano, N. Tsubouchi, "A New Isolation Technology for Bipolar Devices by Low Pressure Selective Silicon Epitaxy," *Tech. Digest of Symposium on VLSI Technology 1982*, pp. 116-117.

[10.11] N. Endo, K. Tanno, A. Ishitani, I. Kurogi, H. Tsuya, "Novel Device Isolation Technology with Selective Epitaxial Growth," *Tech. Digest of IEDM 1982*, pp. 241-244.

[10.12] T. Shibata, R. Nakayama, K. Kurosawa, S. Onga, M. Konaka, and H. Iizuka, "A Simplified Box (Buried Oxide) Isolation Technology for

Megabit Dynamic Memories," *Tech. Digest of IEDM 1983*, pp. 27-30.

[10.13] S.-Y. chiang, K. M. Cham, D. W. Wenocur, A. Hui, and R. D. Rung, "Trench Isolation Technology for MOS Applications," *Proc. of the First International Symposium on VLSI Science and Technology 1982*, pp. 339-346.

[10.14] H.-J. Voss and H. Kurten, "Device Isolation Technology by Selective Low-Pressure Silicon Epitaxy," *Tech. Digest of IEDM 1983*, pp. 35-37.

[10.15] K. Nakamura, M. Yanagisawa, Y. Nio, K. Okamura, and M. Kikuchi, "Buried Isolation Capacitor (BIC) Cell for Megabit CMOS Dynamic RAM," *Tech. Digest of IEDM 1984*, pp. 236-239.

[10.16] J. A. Appels, E. Kooi, M. M. Paffen, J. J. H. Schatorje, and W. H. C. G. Verkuylen, *Phillips Res. Rep.*, 25, 1970, pp. 118.

[10.17] Daeje Chin, private communication.

Chapter 11

A Study of LDD Device Structure Using 2-D Simulations

In this chapter, analysis and design of LDD (Lightly Doped Drain) devices using two-dimensional device simulation and experiments will be described to illustrate the usefulness and necessity of using computer aided design tools in the fabrication of VLSI devices. First, the problem of high electric field in VLSI devices and the use of LDD device as a possible solution is discussed. The fabrication and simulation of LDD device is then described. Finally, the performance, characteristic, physics and design considerations of LDD devices are presented in detail.

11.1 High Electric Field Problem in Submicron MOS Devices

In Very Large Scale Integrated (VLSI) circuits, the dimensions of devices are continually being scaled down to obtain higher density and speed. The channel length, junction depth and gate oxide thickness are scaled down while the channel doping is scaled up. These scalings are necessary to provide more current drive while maintaining the right device threshold voltage and low leakage current. However, the power supply voltage tends to remain unchanged due to system requirements. The same voltage is hence being applied to a much shorter channel which results in a higher channel electric field. In addition, the shallower junction and thinner gate oxide also make the electric field at the drain junction higher. Eq. (11.1a,b) [11.1],[11.2] is a simple analytical expression relating the maximum drain electric field to device parameters.

$$E_m = \frac{(V_D - V_{DSAT})}{\sqrt{3T_{ox}X_j}} \qquad (11.1a)$$

$$V_{DSAT} = \frac{(V_G - V_T)LE_{SAT}}{(V_G - V_T + LE_{SAT})} \qquad (11.1b)$$

where E_m is the maximum drain electric field, V_D is the applied drain voltage, V_{DSAT} is the drain saturation voltage, T_{ox} is the gate oxide thickness, X_j is the junction depth, L is the effective channel length and E_{SAT} is the critical field for velocity saturation and is about 3E4 V/cm.

It can be seen that as gate oxide thickness, junction depth and channel length are scaled down, the maximum drain electric field increases. Electrons in the channel are accelerated by this high drain electric field to reach high energy. These high energy electrons are called 'hot electrons' and they can cause device failure and long term reliability problems.

One problem is that some electrons can reach high enough energy to cross over the silicon/silicon dioxide barrier and create damage at the oxide/silicon interface. The interface damage can cause threshold voltage shift and also reduce the mobility of electrons. The current driving capability of the device is hence continually being degraded during device operation. This presents a severe reliability problem because circuit failure can occur after a certain period of operation. Another problem is that the high energy electrons can cause impact ionization in the drain depletion region, generating electron/hole pairs. The electrons are collected by the drain while the holes go into the substrate. The substrate hole current forward biases the source junction and electrons are injected into the substrate. These electrons will be collected at the drain as excess current and generate more impact ionization substrate current. This is a positive feedback mechanism which can cause the drain current to increase rapidly. The resulting avalanche breakdown will destroy the device. Excessive substrate current can also cause problems to substrate bias generation circuits and is a potential cause of latchup in CMOS circuits. Eq. (11.2) [11.3] is an empirical analytical expression relating substrate current to maximum drain electric field.

$$I_{\text{sub}} \simeq 2I_{DS} \exp(-1.7E6/E_m) \tag{11.2}$$

where I_{DS} is the drain current and E_m is the maximum drain field.

New device structures are needed to reduce the drain electric field and hence prevent hot electron degradation and catastrophic drain breakdown. Several device structures have been proposed, such as lightly doped drain (LDD) device [11.4],[11.5] and AS/P double diffused device [11.6]. The idea is to use a lightly doped region to drop off some drain voltage so that the drain electric field can be lowered. A more graded junction also helps to reduce drain electric field. In the design of such devices, a lot of issues need to be considered. Major concerns are whether the additional n⁻ region will degrade device performance, what is the right doping concentration of the n⁻ region to achieve both low electric field and acceptable series resistance, and whether low electric field leads to less hot electron related problems. In this chapter, only the LDD device will be discussed.

11.2 LDD Device Study

Lightly doped drain (LDD) device has received much attention in recent years as an important VLSI device. The major advantage of LDD device is that the lightly doped region can reduce the peak electric field at the drain. This results in higher breakdown voltage, lower substrate and hot electron currents and hence improved reliability. Other proposed advantages are reduction of short channel effect and improvement in punchthrough voltage due to the shallower tip junction. Experimental results on LDD device have been reported in several papers [11.4],[11.5]. However, the physics of submicron LDD device is not well understood.

This is due to the complexity of two dimensional device effects and difficulties in measuring and extracting parameters for submicron devices. In this section, the device physics for a half micron LDD device is investigated by using the two-dimensional device simulation program CADDET together with experimental results. The combination of computer simulation and experimental analysis is found to provide important insights into the understanding of the LDD device in terms of breakdown, trapping, punchthrough and others. Simulation results are found to agree well with experiments.

Device Fabrication and Simulation

LDD device differs from conventional devices by having a n^- region at both the source and drain side. Fig. 11.1 shows the conventional and LDD device respectively. Different processing methods can be used to create the LDD device structure. In our laboratory, a n^- implant is done after the polysilicon gate is defined and etched. (The n^- implant dose will be in unit of cm^{-2} throughout the chapter.) A thin oxide is grown after the implant and then a layer of oxide from 100 nm to 300 nm is deposited by low pressure chemical vapor deposition (LPCVD) at 900 deg C. This oxide layer is then anisotropically etched to form an oxide spacer at the polysilcion gate edge. Heavy dose Arsenic implant is then done to create the n^+ source and drain junctions. Fig. 11.2 shows the processing sequence used to fabricate the LDD device. The device used in this work has gate oxide thickness of about 20 nm, a junction depth of about 0.25 micron and a n^- region with various doping levels and lengths.

The two-dimensional simulation program CADDET used to study the LDD device solves both the Poisson equation and the current continuity equation simultaneously. The program does not have an impact ionization model and hence cannot simulate substrate current. However, since the substrate current is related to the maximum drain electric field (Eq. (11.2)), very useful information can still be obtained. In CADDET, the n^- region can only be placed either on the source or drain end. The

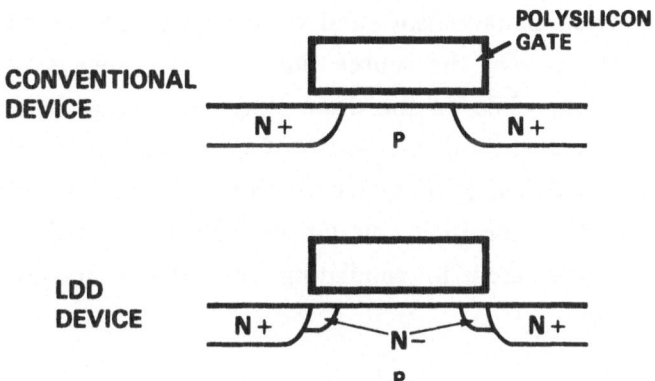

Fig. 11.1 Conventional and LDD device structures.

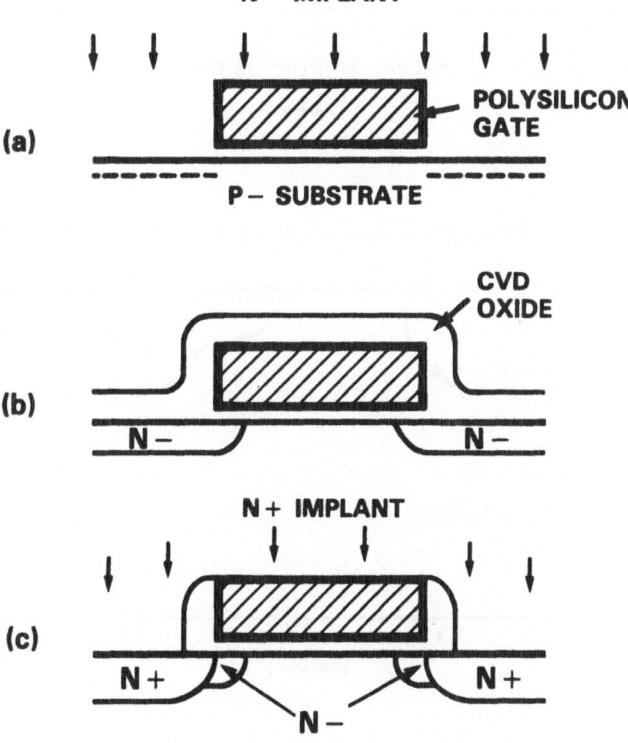

Fig. 11.2 Processing sequence for LDD device.

tip is hence placed on the source side to study the effect of series resistance on device characteristics and on the drain side to study drain electric field effects. On the source side, the gate always covers the entire source region, while on the drain side, the overlap between gate and drain can be varied. Both the length of the n⁻ region and its doping profile can be varied. The particular limitations of CADDET need to be taken into account in the interpretation of simulation results. Fig. 11.3 shows the structure used in simulating the source and drain side respectively.

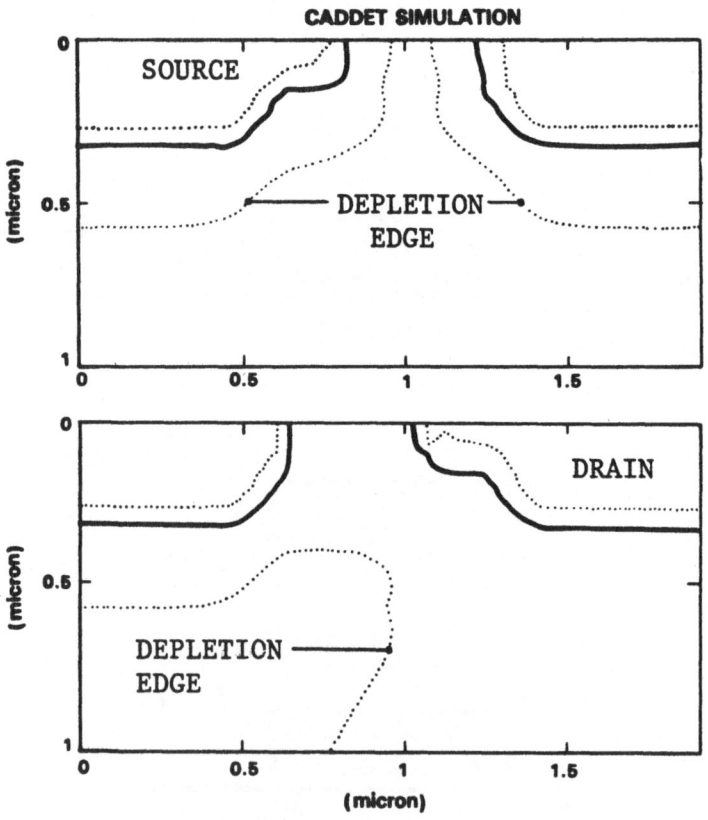

Fig. 11.3 Structures used for source and drain simulations

Performance Comparison Between LDD and Conventional Device

One common criteria used to evaluate a certain device technology is to plot the saturation current versus the punchthrough voltage for the devices. The idea is that the shorter the channel length, the higher is the current but at the same time, the lower is the punchthrough voltage. For the same channel length, a technology with a deeper junction will have lower punchthrough voltage while one with lower source/drain doping and shallower junction depth (hence higher series resistance), will have lower saturation current. The saturation current versus punchthrough voltage characteristic tells us the best current drive we can get subject to a certain punchthrough voltage requirement and is hence a good evaluation of the performance of a technology. In simulating the performance comparison between LDD and conventional device, a 0.2 micron n^- tip is placed at the source side. The n^- dose and channel lengths are varied. For punchthrough simulation, the gate voltage is set at zero volt and the drain voltage is stepped in 0.1 volt increments. The punchthrough voltage is defined as the drain voltage at which the drain current is 10 nA per micron channel width. Saturation current is defined at $V_G = V_D = 5$ volts. Fig. 11.4 shows a plot of the simulated saturation current versus punchthrough voltage characteristic. It can be seen that for the same punchthrough voltage, LDD device is about the same or lower than conventional devices, depending on the dose of the n^- region. Fig. 11.5 is a plot of I_{DSAT} versus punchthrough characteristic from some experimental LDD and conventional devices, which agrees with the simulation. The reason for such inferior performance in the half micron device is due to the fact that the series resistance of the n^- region is quite comparable to the channel resistance. The gain in punchthrough voltage due to the shallow n^- junction is more than offset by the decrease in current due to the series resistance. Fig. 11.6 and 11.7 plots the simulated linear and saturation transconductance characteristic for conventional and LDD devices under different gate bias. The numbers in parenthesis are the effective channel lengths. In the linear region, the percentage

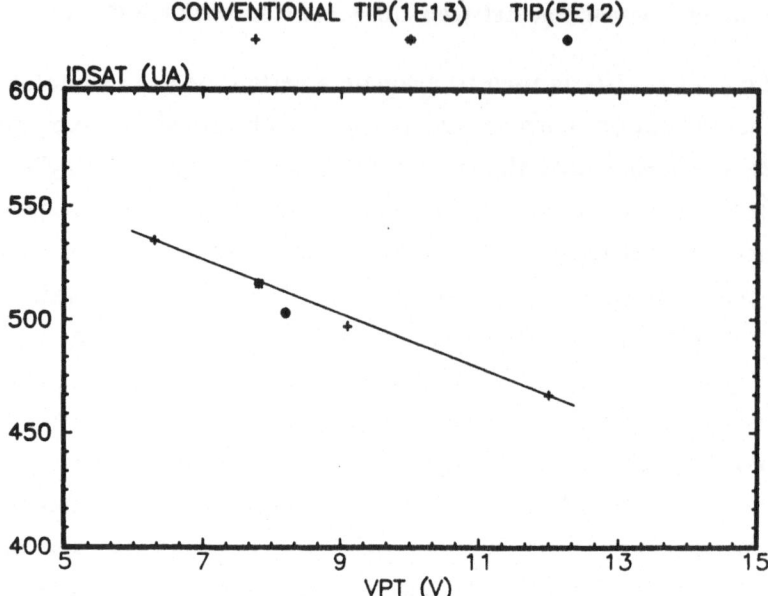

Fig. 11.4 Saturation current vs. punchthrough voltage,
simulation. Tip length = 0.2 μm.

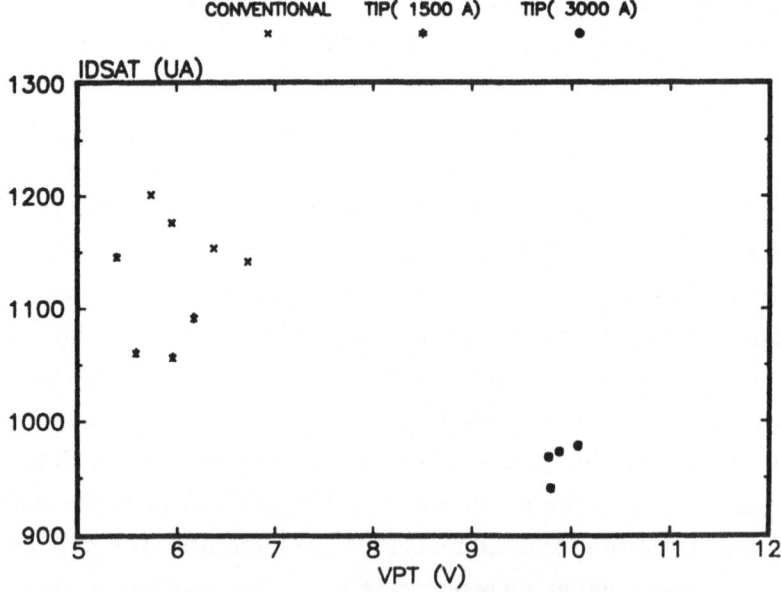

Fig. 11.5 Saturation current vs. punchthrough voltage,
experimental data. Tip dose = 5E12 cm^{-2}.

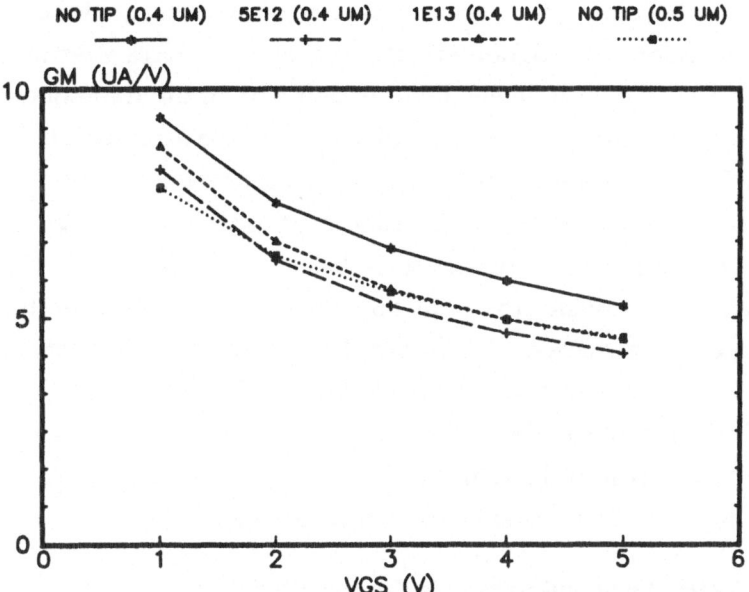

Fig. 11.6 Simulated linear transconductance vs. gate bias. Tip length = 0.2 μm.

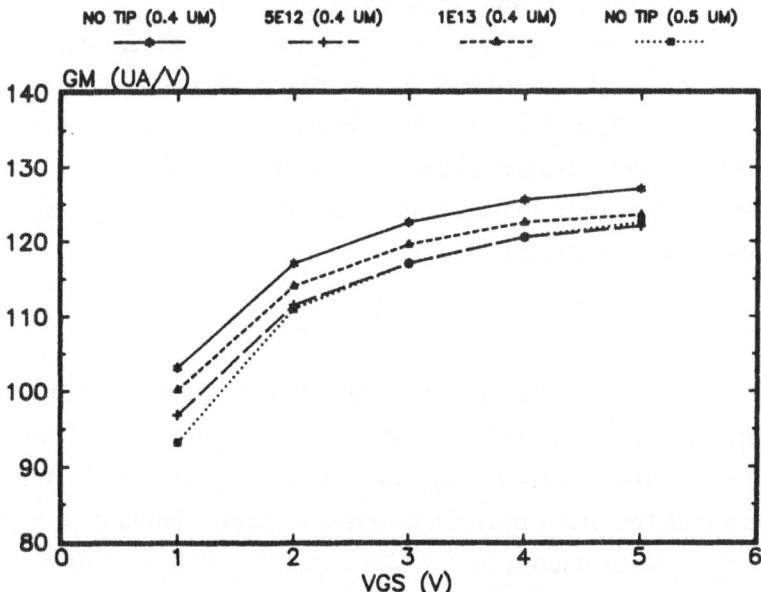

Fig. 11.7 Simulated saturation transconductance vs. gate bias. Tip length = 0.2 μm.

degradation becomes worse with increasing gate voltage, whereas in the saturation region, degradation effect is less and decreases with increasing gate voltage. The experimental linear and saturation transconductance characteristics shown in Fig. 11.8 and 11.9 show the same trend. Since in CADDET, the n^- region on the source side is entirely covered by the gate, the effective series resistance is reduced at high gate voltage and the degradation due to this resistance is being under estimated. However, CADDET does provide the right direction and saves a lot of time and effort in experimentation. The typical time required to fabricate and test the devices is about 2-3 months while the simulations can be done on a HP1000 minicomputer in a few days. Moreover, the physics of the effects can be readily investigated by looking at potential, electric field and electron distributions in the simulated device.

Drain Electric Field and Substrate Current Study

The drain electric field characteristic of LDD device is simulated by putting the n^- region at the drain side. The drain electric field as a function of n^- region doping, n^- region length and the amount of gate overlap of the n^- region is studied. Several interesting results are obtained through the simulation and are confirmed experimentally by looking at the substrate current characteristics. Important physical insight into the operation and design of LDD device, conventional device and minimum overlap devices are obtained.

A.) *Drain electric field as a function of n^- region dose*

In the simulation of drain electric field characteristic as a function of n^- region doping, it is found that contrary to common belief, lower n^- region doping doesn't necessarily lead to lower drain field. Fig. 11.10 plots the simulated drain peak field versus n^- region implant dose for n^- region length of 0.2 micron and a bias condition of 5 volts at the drain and 3 volts at the gate. A minimum peak field is found at a dose of 5E12. Physically, for the lowly doped case (2E12), the entire n^- region is

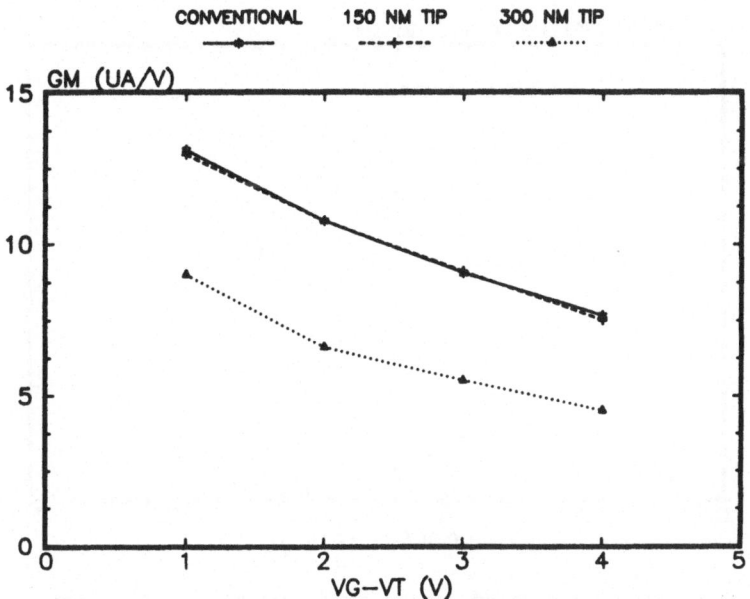

Fig. 11.8 Experimental linear transconductance vs. gate bias. Tip dose = 5E12 cm^{-2}.

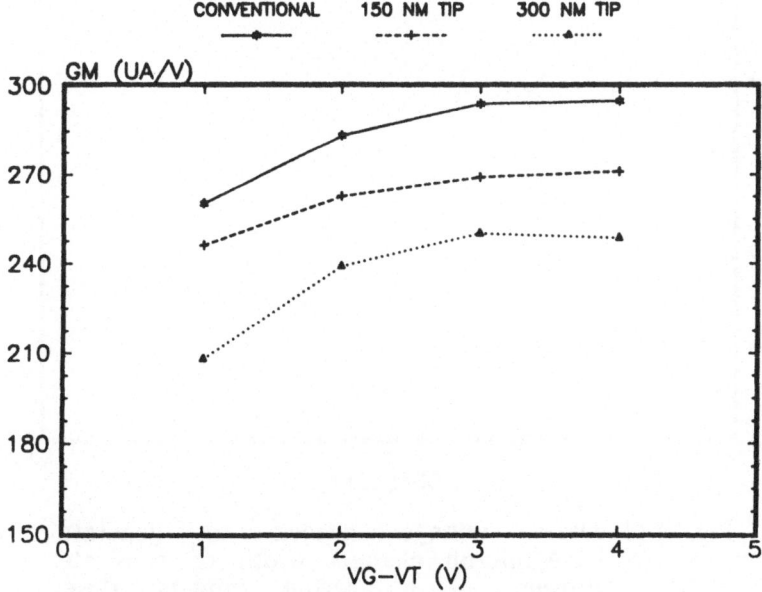

Fig. 11.9 Experimental saturation transconductance vs. gate bias. Tip dose = 5E12 cm^{-2}.

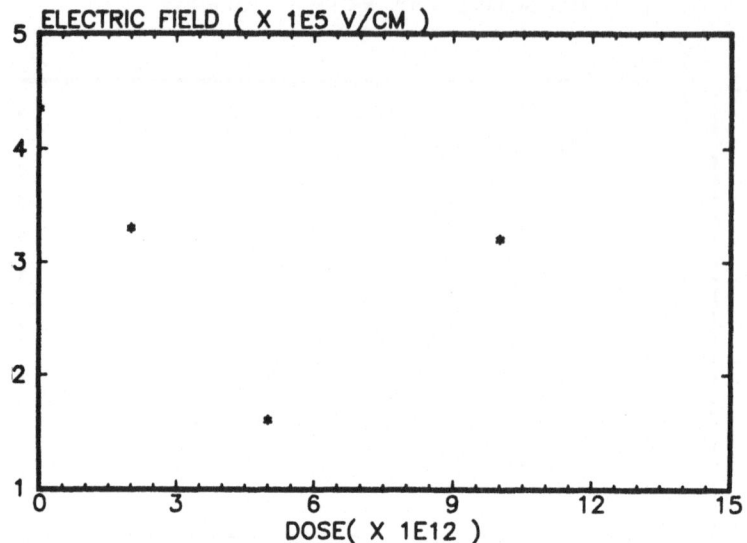

Fig. 11.10 Simulated drain peak field vs. n⁻ region
implant dose. Tip length = 0.2 μm.

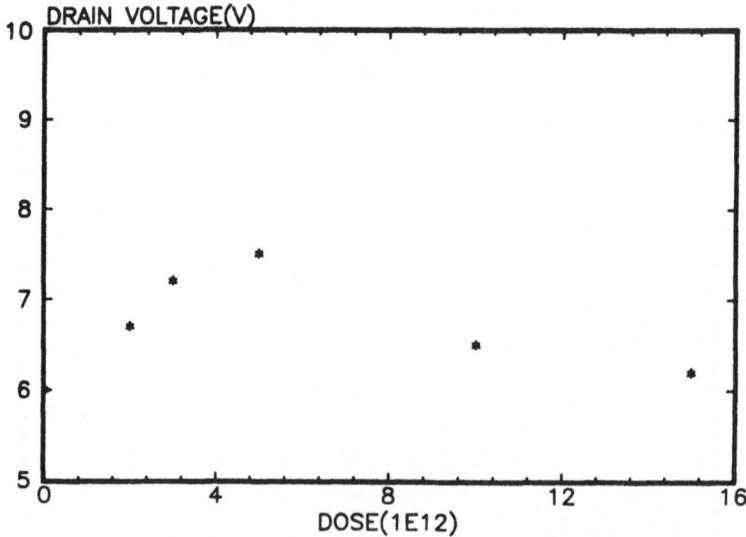

Fig. 11.11 Drain   voltage   needed   to   generate
1 μA/micron channel width of substrate
current   vs.   n⁻   region   implant   dose,
experimental data. Tip length = 0.2 μm.

depleted and a high field occurs at the high-low junction. For the higher doping (1E13), it is harder to deplete the n⁻ region and the peak field occurs at the edge of the n⁻ region. Experimental results confirm this finding. Fig. 11.11 plots the drain voltage needed to generate 1 μA per micron of channel width of substrate current versus the n⁻ region implant dose. It can be seen that the highest drain voltage required is at a dose of 5E12. This optimum dose for breakdown and lowest substrate current will vary to a certain extent with the n⁻ region length. Although 5E12 gives the best breakdown and substrate current characteristics, the series resistance is high as shown in figures 11.4 to 11.9. Another consideration is that lower substrate current does not necessarily lead to lower hot electron trapping. All these factors need to be considered in the optimum design of the device.

B.) *Drain electric field versus gate voltage characteristics*

According to Eq. (11.1a), in the conventional device the maximum drain electric field is directly proportional to $(V_D - V_{DSAT})$. Since V_{DSAT} increases with V_G (Eq. (11.1b)), the drain electric field decreases with increasing gate voltage. From Eq. (11.2), reduction in drain electric field means the substrate current will also decrease with increasing gate voltage. Fig. 11.12 shows the substrate current versus gate voltage characteristic of a conventional device. It can be seen that after an initial peak, the substrate current does decrease with increasing gate voltage. The initial increase of substrate current with gate voltage is due to increase in channel current.

The drain electric field versus gate voltage characteristic is simulated for LDD structures with different n⁻ region doping and amount of overlap between the gate and the n⁻ region. The electric field at the surface along the channel is plotted for each bias point and the position where the peak electric field occurs is noted. Fig. 11.13 shows a plot of the simulated electric field along the channel for the bias point of $V_G = 3$ V and $V_D = 6$ V. The electron concentration as a function of Y

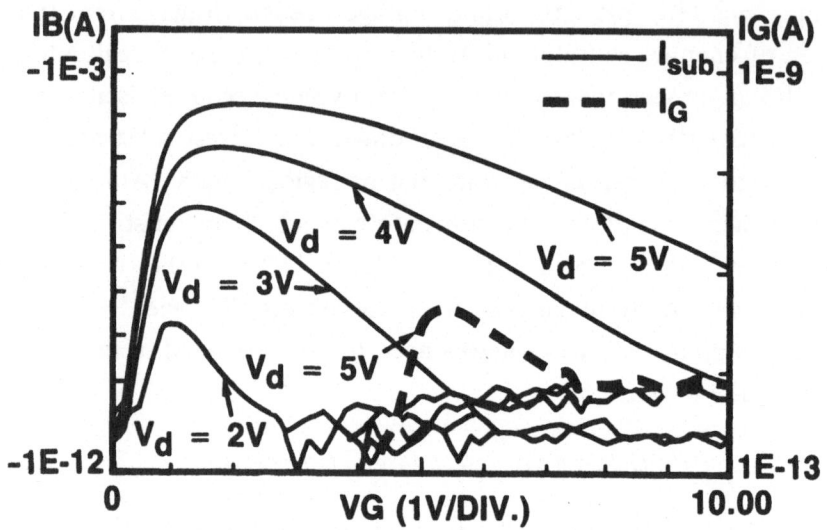

Fig. 11.12 Substrate current vs. gate bias for conventional device.

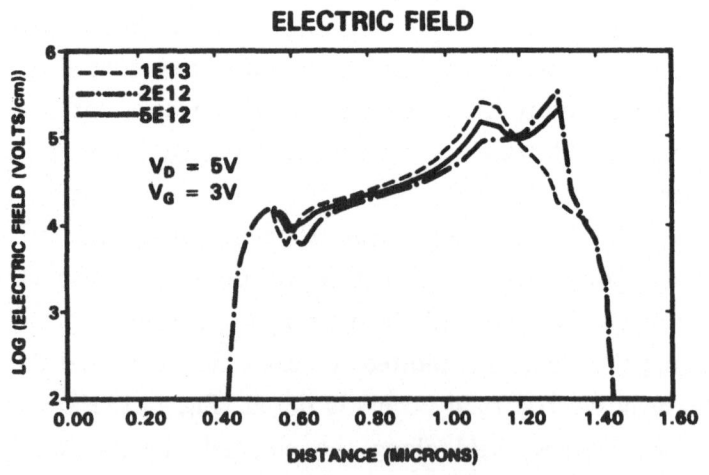

Fig. 11.13 Simulated electric field along the channel.

(the distance from silicon/silicon dioxide interface into silicon) is then plotted at the position of the channel where peak electric field occurs. The position where peak electron concentration occurs is noted. Depending on the device structure, the peak electron concentration in some cases is not at the surface. The electric field along the channel is then plotted again at the Y position where peak electron concentration occurs. Then the maximum drain electric field is found from this plot. Fig. 11.14 is a plot of maximum drain electric field as a function of gate

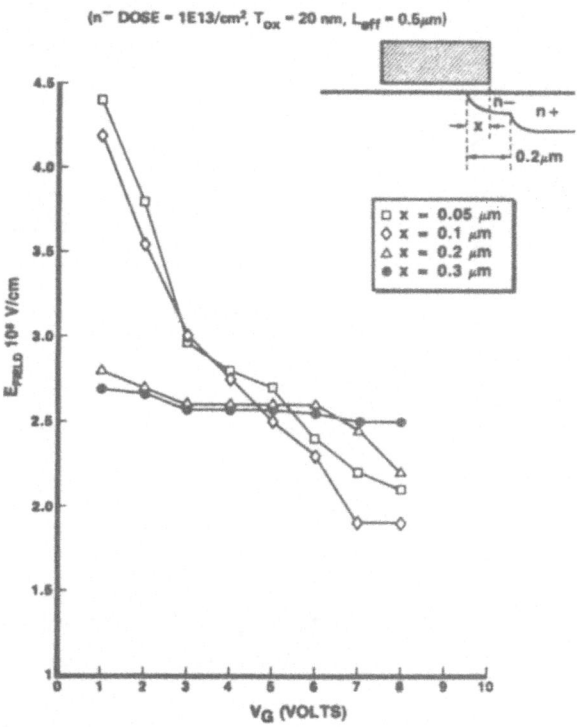

Fig. 11.14 Maximum drain electric field vs. gate bias
for n⁻ dose of 1E13.

bias for different amount of gate overlap at a n⁻ dose of 1E13. It can be seen that for low V_G, the drain electric field is greatly reduced when the gate overlaps the n⁻ region more. Another observation is that the drain electric field is actually not a very sensitive function of gate voltage for half micron LDD device for X larger or equal to 0.2 μm. Fig. 11.15 is a similar plot for a n⁻ region dose of 5E12. The characteristic is very different from that of Fig. 11.13. It can be seen that the electric field initially decreases with increasing gate voltage, reaches a minimum and then goes up again. The smaller the gap between the gate and the n⁺

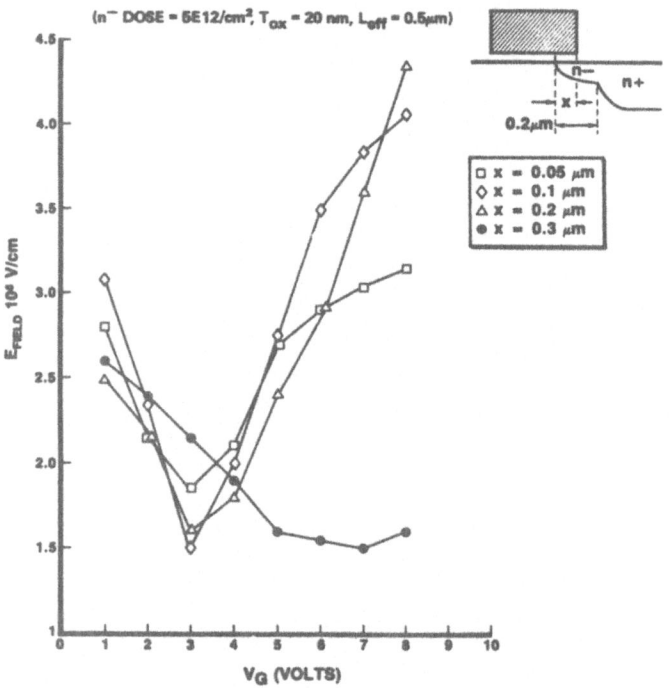

Fig. 11.15 Maximum drain electric field vs. gate bias
for n⁻ dose of 5E12.

junction, the higher the electric field goes up. This effect was confirmed in experimental devices. Fig. 11.16 show substrate and gate current versus gate voltage characteristics for   LDD devices with a n⁻ region doping of

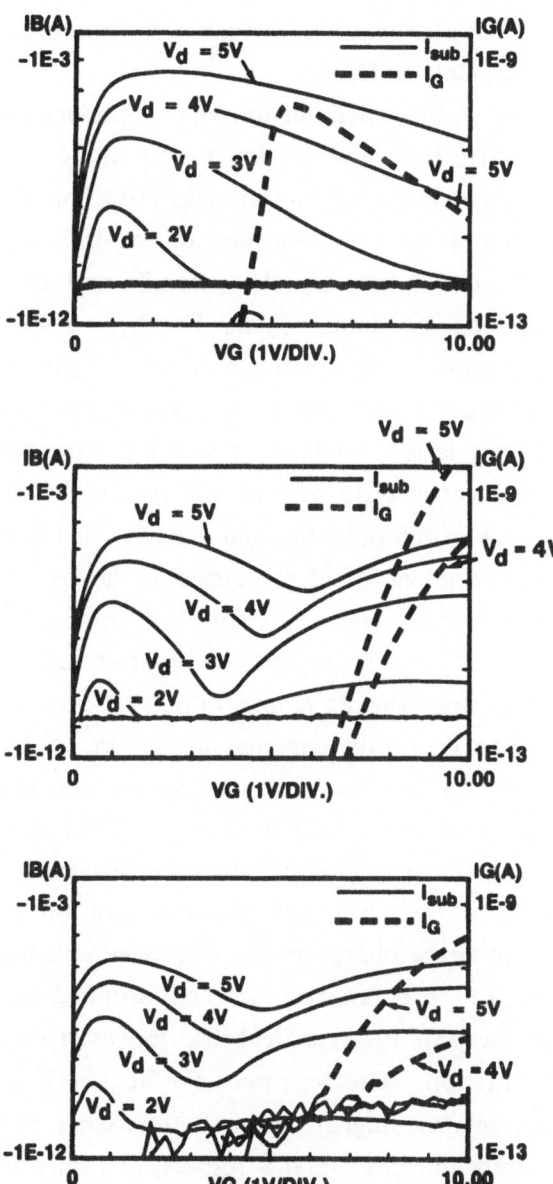

Fig. 11.16 Substrate and gate current vs. gate bias for
LDD devices with different spacer widths.
Top: 120 nm, Middle: 200 nm, Bottom:
300 nm.   (All   with   n⁻   doping = 5E12,
T_{ox} = 20 nm, W/L = 100/1)

5E12 and different n⁻ region length. A double hump is observed in the substrate current characteristics which correspond to a decreasing and then increasing electric field.

The physics of this phenomenon can be understood by looking at the potential, electric field and carrier density along the channel. Fig. 11.17 plots the channel potential and electric field for $V_G = 2$ and 8 volts. It can be seen that at $V_G = 2$ volts, most of the drain potential is dropped in the n⁻ region overlapped by the gate (X) and the peak electric field occurs at the edge of the n⁻ region. For $V_G = 8$ volts, most of the drain potential is dropped in the region between the gate edge and the edge of the n⁺ junction (Y) and the peak electric field occurs at the Hi-Lo junction. In the latter case $(V_G = 8$ volts), the shorter the distance Y, the higher is the electric field because the potential is dropped in a shorter distance. The way the potential and electric field behave under different gate bias conditions is found to be due to the fact that at high gate bias the channel carrier density can be comparable to the doping concentration of the n⁻ region. At $V_G = 2$ volts, the channel carrier density is less than the n⁻ region doping concentration. The n⁻ region under the gate is mostly depleted and most of the drain potential is dropped there. The peak electric field occurs at the edge of the n⁻ region. However, at $V_G = 8$ volts, the channel carrier density is comparable to the n⁻ region doping concentration (for the 5E12 case), therefore the n⁻ region under the gate can no longer be considered depleted. The space charge region is pushed towards the heavily doped n⁺ junction and the potential is then dropped in the region Y. The position of the peak electric field also moves from the p/n⁻ junction to the n⁻/n⁺ junction. The narrower the region Y is, or the higher the channel current, the higher is the drain electric field which results in higher substrate current. In this respect, this phenomenon resembles the Kirk effect in bipolar transistors. Since the effect is due to channel carrier density being comparable to n⁻ region doping concentration, increasing n⁻ region doping will suppress this effect while reducing gate oxide thickness will enhance it. This double hump substrate current

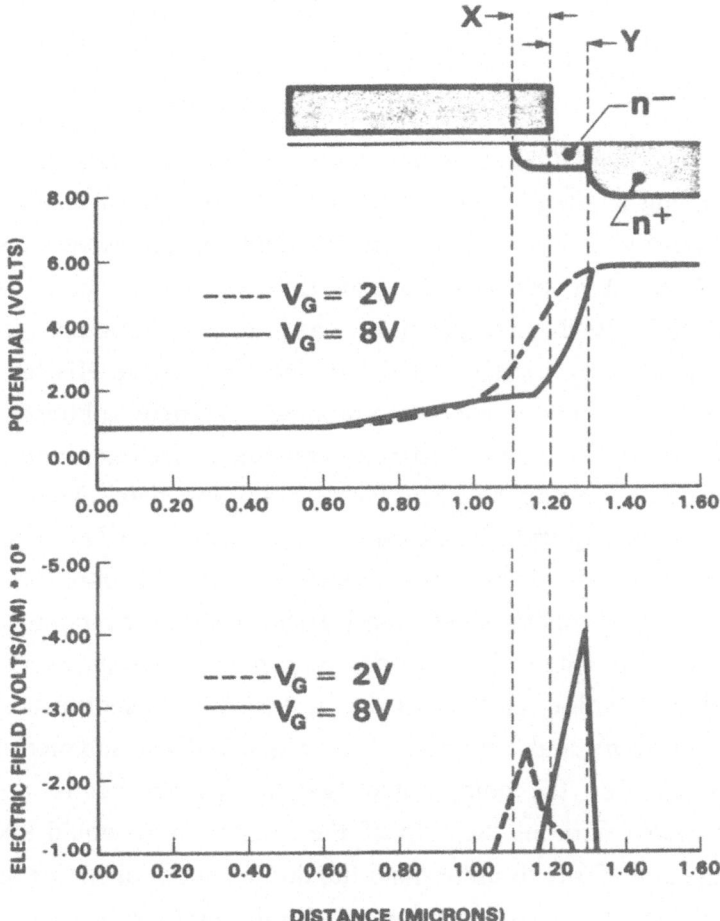

Fig. 11.17 Channel potential and electric field along
the channel for $V_G = 2$ and 8 V.

characteristic is found in minimum overlap devices. When there is not enough overlap between the n^+ source/drain and the gate, a graded n^- region exists between the edge of the gate and the n^+ region and this phenomenon can be observed. This characteristic can hence be used as a tool to check if there is enough overlap between the source/drain and the gate. Adequate overlap between the source/drain and gate is important both for device performance and reliability.

11.3 Summary

Computer simulation together with experiments provide a good understanding of the physics of LDD devices. Submicron LDD device is found to provide lower drain electric field at the expense of lower current drive. A n^- region doping of 5E12 is found to give the lowest drain electric field although this may not be optimum from the standpoint of series resistance and hot electron damage effects. A dose of 1E13 is probably a good compromise. A new substrate current characteristic is found and explained. This can be used as a tool to check for inadequate overlap between source/drain and gate. These physical understandings help us to be aware of the various tradeoffs in the design of an optimum LDD device for a certain circuit application.

In this study, computer simulation is found to have several advantages. First of all, the physics of the device can be investigated in detail. Potential distribution, electric field and electron concentrations can be plotted out easily and analyzed. This is essential for submicron device study because of the complicated two-dimensional nature and other subtle effects. Without the help of the simulation, it would take much more time and effort to understand the double hump phenomenon in the substrate current characteristic. Another important feature is the ease of varying different device parameters. This saves a lot of time and money and at the same time eliminates uncertainties in experiments. The device structure is well defined in terms of junction doping profile, channel length and channel doping profile. However, a simulator is only as good as the physical models in it. One must be aware of the limitations in the physical models, device structures and other specifications in the model and interpret the results properly. The simulation results need to be confirmed with experiments. Simulators can be used to guide us in the direction of more intelligent experiments and greatly reduce the number of experiments needed. Experimental results can also be interpretated with the help of simulations. This study on LDD devices shows that proper use of simulation with experiment can be very effective in the understanding and design of submicron devices.

References

[11.1] Cheming Hu, "Hot electron effects in MOSFETs", *Tech. Digest of IEDM 1983*, pp. 176-179.

[11.2] P. K. Ko, Ph. D thesis, University of California, Berkeley, CA 1982.

[11.3] S. Tam et. al., *IEEE Trans. Electron Devices*, ED-29, pp. 1740-1744, Nov 1982.

[11.4] P. J. Tsang, S. Ogura, W. W. Walker, J. F. Shepard and D. L. Critchlow, "Fabrication of High performance LDDFET's with oxide sidewall-spacer technology," *IEEE Trans. on Electron Devices*, ED-29, no.4 pp. 590-595, April 1982.

[11.5] Seiki Ogura, Christopher F. Codella, Nivo Rovedo, Joseph F. Shepard and Jacob Riseman, "A half micron MOSFET using double implanted LDD," *Technical Digest of IEDM 1982*, pp. 718-721.

[11.6] K. Balasubramanyam, et al, "Characterization of As-P Double Diffused Drain Structure," *Tech. Digest of IEDM 1984*, pp. 782-785.

Chapter 12
MOSFET Scaling by CADDET

12.1 Introduction

The microcircuit density and performance have been increased by many orders of magnitude during the era of the integrated circuits (I.C.) through the process of scaling. As pointed out in the overview chapter, the long channel relations are not strictly valid for horizontal dimensions that are comparable to the vertical dimensions. In this example, the operating voltage is kept constant. The horizontal dimension, L_{eff}, and the vertical dimension, T_{ox}, will be separately scaled to approximately two-thirds of the established process values. The two scaling factors are not identical, but are in the typical scaling range. The result of this reduction is then established. Comparison with the long-channel scaling assumptions is possible, and the importance of secondary physical effects can be seen. Device width will not be scaled, the current drive capability will be expressed for a fixed width. Breakdown and punchthrough voltages must be sufficiently greater than the supply voltage so that reliability is not a problem. For this example, power density is not a limitation.

Two different modes of devices are considered, an enhancement mode device and a depletion mode device. Section 12.2 discusses how device simulation programs are used to study an enhancement mode device scaling and its results are compared to the classical predictions [12.1].

Next, a scaling scheme of a depletion mode device is considered in Section 12.3 and is contrasted to the enhancement mode device scaling. Section 12.4 presents conclusions. The device simulation program CADDET is used throughout for numerical calculations.

12.2 Scaling of an Enhancement Mode MOSFET

CADDET Simulation

A typical n-channel enhancement mode device is chosen as a standard device to develop the discussion. The mask dimensions of the gate are 50 μm wide and 2 μm long. Structural data are collected by a series of measurements. The gate oxide thickness is 40 nm measured by C-V technique. [12.2] The measured effective channel length (L_{eff}) and the source-drain series resistance are 1.25 μm and 750 Ω per μm width respectively [12.3]. The electrical characteristics of the device are shown in Fig. 12.1. The measured data are corrected for the voltage drop by the source-drain contact resistances.

The electrical characteristics of short channel length devices deviate from the classical model and do not conform to any known simple analytic expressions [12.4]. Numerical process and device simulation programs become very useful tools to study the characteristics of the small geometry devices due to the built-in generality in the device structure and the accuracy in numerical computation.

CADDET is a 2-D device characteristics simulation program developed by HITACHI. The program solves Poisson's equation in the substrate with the current continuity equation for a single carrier only. It assumes that the surface of the substrate is planar and that the gate covers the entire substrate surface. In our application, the electron mobility in the program is replaced by a new expression given by Eq. (3.15). The substrate doping profile is simulated by running SUPREM [12.5], a 1-D process simulation program, which gives a doping profile in the substrate when the process sequences are specified. Since CADDET takes only

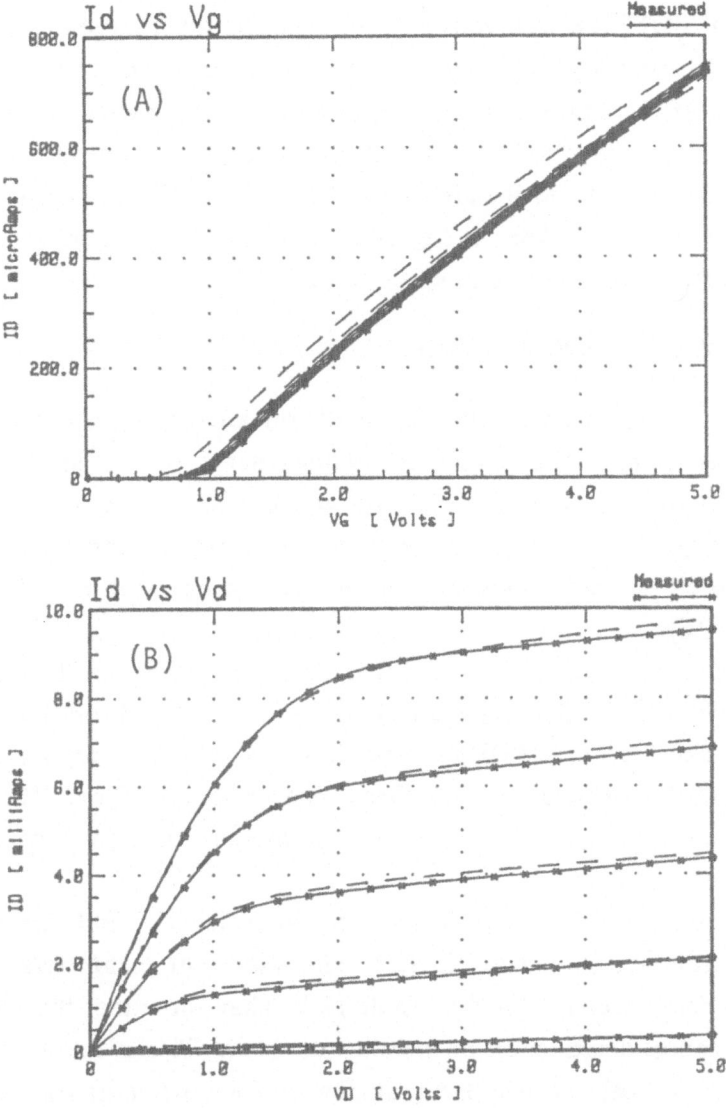

Fig. 12.1 Comparisons between the measurements (solid lines) and CADDET simulations (dashed lines). (A): $V_B = 0$ to -5 V, (B): $V_G = 1$ to 5 V. Device: $T_{ox} = 40$ nm, $L_{eff} = 1.25$ μm, $W_{eff} = 48.25$ μm

analytic expressions (such as normal distribution) for substrate doping, the results of SUPREM are converted to a normal distribution as shown in Table 12.1. Impurity distribution at the source or drain to substrate are obtained as in the discussion of SUPREM (Ch. 2). Based on these data, the current-voltage (*I-V*) characteristics of the present (unscaled) enhancement mode device are simulated by CADDET and are compared with the measured data in Fig. 12.1. The device parameters that are important for digital circuits are listed in Table 12.2.

Scaling of Gate Oxide Thickness

As a first step in scaling, the gate oxide thickness (T_{ox}) is reduced from 40 nm to 25 nm. This gives the scaling factor of 1.6 in T_{ox}. The reduction in T_{ox} increases the gate capacitance (C_{ox}), which in turn increases the transconductivity (g_m) and the drain current. According to the classical model, the drain current is proportional to this scaling factor [12.6].

The threshold voltage (V_T) depends on the implantation dosage to the first order when the depletion boundary is formed outside of the implantation region. CADDET simulations are done to study V_T shift by the implantation dosage with fixed energy of 50 KeV. The boron implantation dosage of $1.4E12/cm^2$ is chosen to keep V_{T0} (V_T at $V_B = 0$ V) at 0.62 V.

I-V characteristics of the device with the scaled T_{ox} and the adjusted implantation are shown in Fig. 12.2. The important parameters of these characteristics are also listed in Table 12.2. They show that the reduction in T_{ox} causes a reduction in mobility of 14 %. This is due to the fact that the vertical electric field at the surface degrades the electron mobility. With a fixed gate bias, the thinner the insulator, the higher the electric field at the surface and the smaller the surface electron mobility. Fig. 12.2 also shows that the saturation current at $V_{GS} = V_{DS} = 5.0$ V is 13 mA, increased from 9.2 mA by factor of 1.4. This number is close to

SUPREM Results

	Channel Implant	Junction Implant	Depletion Implant
Element	Boron	Arsenic	Arsenic
Dosage[$/cm^3$]	6.0E11	5.0E15	1.3E12
N_B[$/cm^3$]	1.0E15	1.0E15	1.0E15
Peak Concentration [$/cm^3$]	3.1E16	3.3E20	1.1E17
Peak Position [μm]	0.12	0.00	0.07

Gaussian Parameters

	Channel Implant	Junction Implant	Depletion Implant
Dosage[$/cm^3$]	6.0E11	5.0E15	1.3E12
Rp [μm]	0.12	0.00	0.07
ΔRp [μm]	0.0825	0.0600	0.0429

Table 12.1 Conversion of SUPREM results to Gaussian formula.

W/L [μm/μm]	50/2	50/2	50/1.5
T_{ox} [A]	400	250	250
μ0 [cm^2/V/Sec]	638	552	552
V_{TO} [V]	0.674	0.619	0.783
N_B [/cm^3]	3.44E14	3.48E14	3.01E14
R_S [Ω]	15	15	15
R_D [Ω]	15	15	15
ΔL [μm]	0.375	0.375	0.375
ΔW [μm]	0.625	0.625	0.625

Table 12.2 SPICE parameters for unscaled and scaled
enhancement mode devices

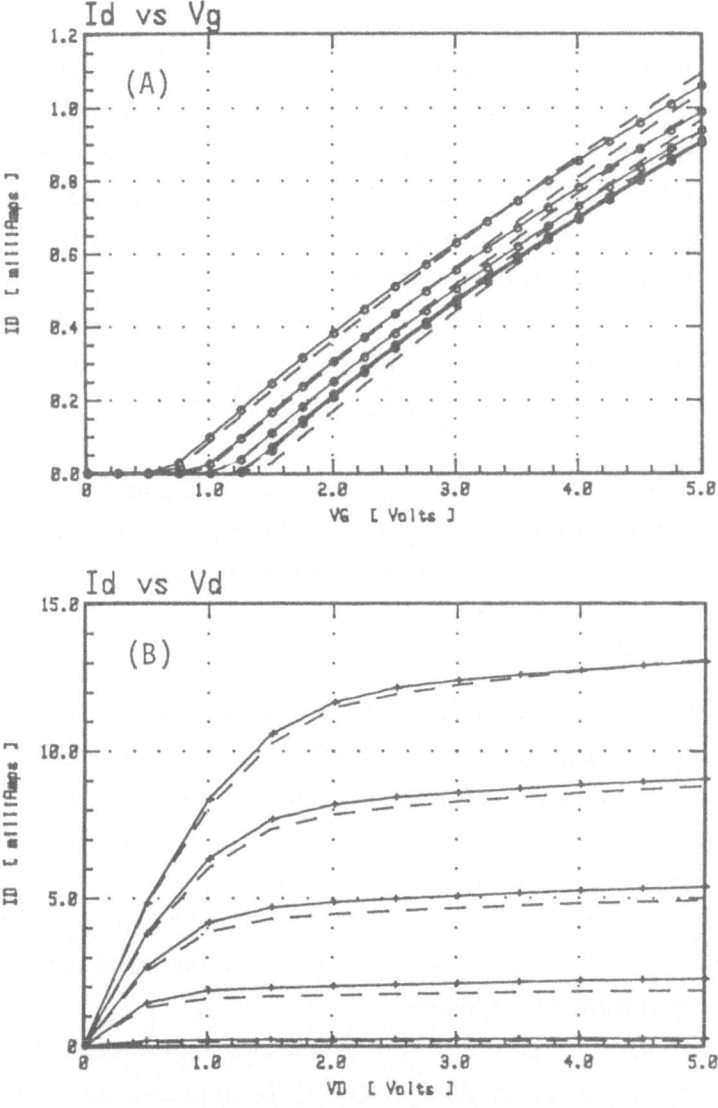

Fig. 12.2 Simulations of an enhancement mode device with T_{ox} = 25 nm, W_{eff} = 48.25 μm and L_{eff} = 1.25 μm. The solid lines are by CADDET and dashed lines by SPICE. (A): V_B = 0 to -2 V; (B): V_G = 1 to 5 V.

the product of the scaling factor of T_{ox} (= 1.6) and the reduction of mobility (= 0.86).

It should be noted that the increase in drain current at this stage of scaling does not always lead to an increase in circuit speed. The premise is true when the load of an invertor stage is dominated by parasitic capacitances, thus the increase in the gate capacitance does not affect the total load capacitance very much. So, with the fixed load, the charging time is inversely proportional to the driving current. However, in case of a circuit whose load is dominated by the gate capacitance, the scaling in T_{ox} alone slows down the circuit speed, because the drain current increases by factor of only 1.4 while the load capacitance increases by 1.6 in our case. Thus the load increases by the higher factor than the drain current.

Since the present scaling in T_{ox} should be followed by the scaling in channel length, which requires the readjustment of implantation dosage and energy, the punchthrough effect is not considered at this moment.

Scaling of Channel Length

With the gate oxide thickness scaled as discussed in the previous section, the channel length in the mask dimension (L) is reduced from 2.0 μm to 1.5 μm. Assuming that the source/drain contacts are made by the same process technology before and after the scaling, this scaling reduces the effective channel length (L_{eff}) from 1.25 μm to 0.75 μm. This results in a scaling factor of 1.67. Considering that L_{eff} is less than one micron, the effects of reducing L_{eff} on the threshold voltage and on the channel current are expected.

With the scaled structure (T_{ox} = 25 nm and L_{eff} = 0.75 μm), the shift of V_{T0} by implantation dosage (Q_{NA}) is simulated at 50 KeV. The implantation dosage of 1.17E12/cm^2 is chosen to get V_{T0} of 0.67 V. With this fixed implantation dosage, the implantation energy is varied to study the subthreshold current and the punchthrough voltage. First, the subthreshold currents (I_{DS} at V_{DS} = 0.1 V, V_{BS} = -2 V) are simulated in

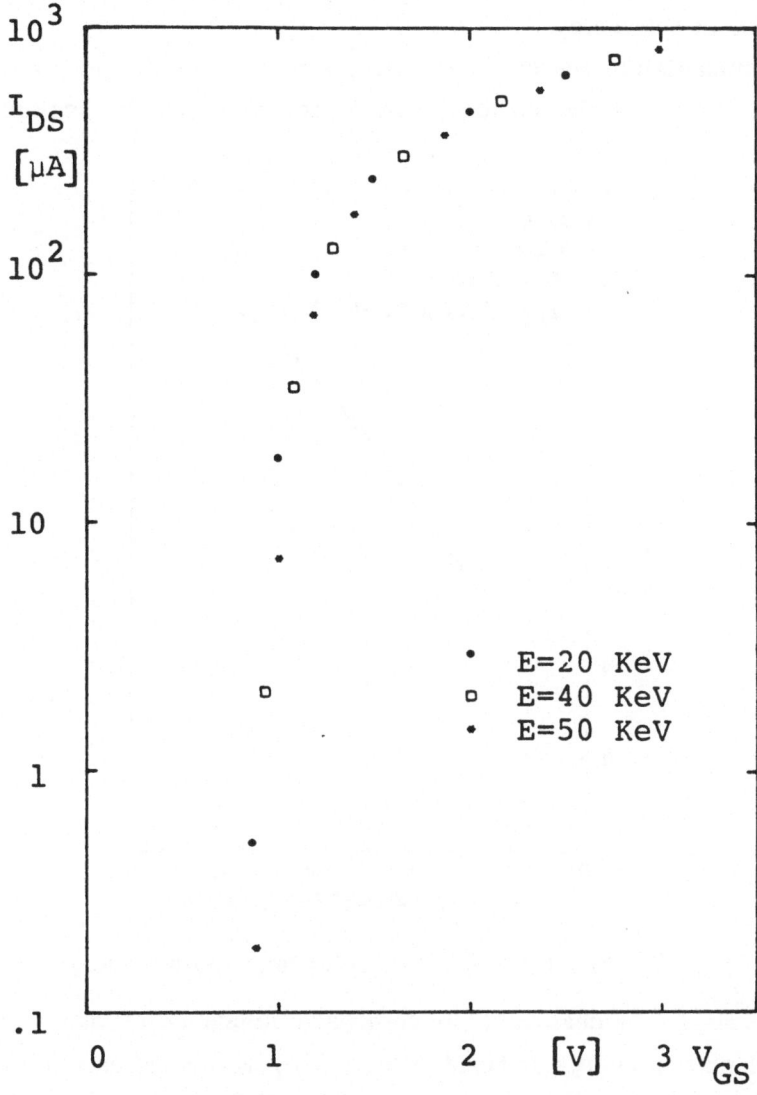

Fig. 12.3 The subthreshold currents of the scaled enhancement mode device. W_{eff} = 48.25 μm, L_{eff} = 0.75 μm, T_{ox} = 25 nm, V_{DS} = 0.1 V, V_{BS} = -2 V, Q_{NA} = 1.17E12 cm^{-2}.

Fig. 12.3 with implantation energy ranging from 20 KeV to 50 KeV. The figure shows that the slope of the current is fixed to 74 mV/decade regardless of the energy.

The implantation energy determines the position of the peak impurity concentration and the current path at the specified bias condition.

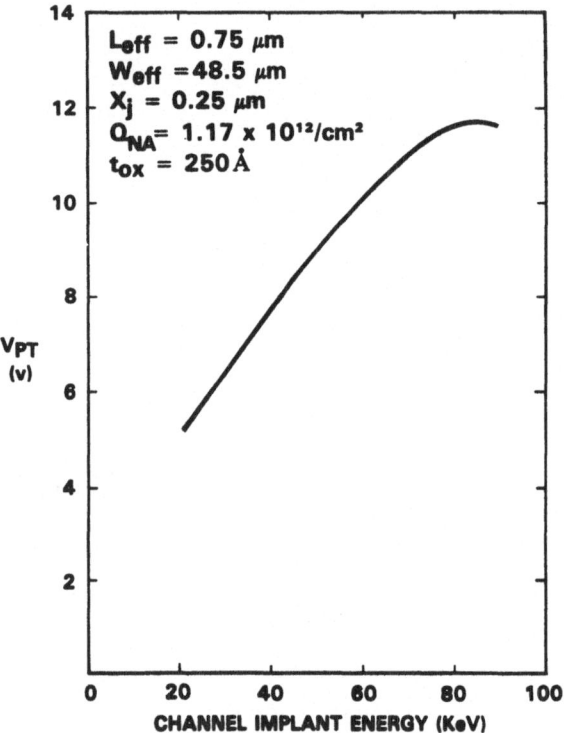

Fig. 12.4 Punchthrough voltage vs. implantation energy.

Fig. 12.4 shows the change of punchthrough voltage (V_{PT}) as a function of implantation energy at fixed dosage. V_{PT} is defined as a drain bias voltage at which the drain current is 1 nA per 1 μm width at $V_{GS} = 0$ V and $V_{BS} = -2$ V. The figure shows that V_{PT} is sensitive to energy and has a maximum value at 80 KeV. Above this value, V_{PT} deteriorates due to the occurrence of surface punchthrough. Below this value, bulk punchthrough is a dominant current mechanism. It is known that the punchthrough voltage is very sensitive to the doping profile in the substrate. Our measurement shows that the boron implantation follows the Pearson IV distribution rather than the normal distribution. If CADDET were upgraded to take different types of impurity distributions, the accuracy of the punchthrough simulation would improve.

The *I-V* characteristics of the scaled structure is shown in Fig. 12.5. The device parameters are extracted from these simulated curves and tabulated in Table 12.2 along with the previous results. The increase of implantation energy to 80 KeV increases V_{TO} to 0.78 V from its target value of 0.62 V at 50 KeV. The high implantation energy pushes the peak of the impurity profile deeper from the surface and the voltage drop across the depletion region increases. Then the threshold voltage, which is the sum of the voltage drops in the insulator and in the depletion region, increases. The table shows that the surface mobility does not change very much by channel length scaling. The shorter channel length promotes the sharing of the depletion charge between the gate terminal and the source/drain terminals [12.7]. This causes a decrease in the effective doping of the short channel device. Fig. 12.5 shows that the drain current at $V_{DS} = V_{GS} = 5$ V is 18 mA, increased from 13 mA of the unscaled channel length by factor of 1.39, which is less than the scaling factor of channel length (1.67). Considering that the current increase at $V_{GS} = 5$ V and $V_{DS} = 0.1$ V is from 0.9 mA to 1.6 mA, a factor of 1.78, the drain current increase by factor of only 1.39 in the saturation region is considered to be due to the carrier velocity saturation in the channel. The resultant gain of the drain current by two consecutive scalings is 1.96, far less than the product of two scaling factors (2.67) as predicted by the classical model.

12.3 Scaling of a Depletion Mode MOSFET

CADDET Simulation

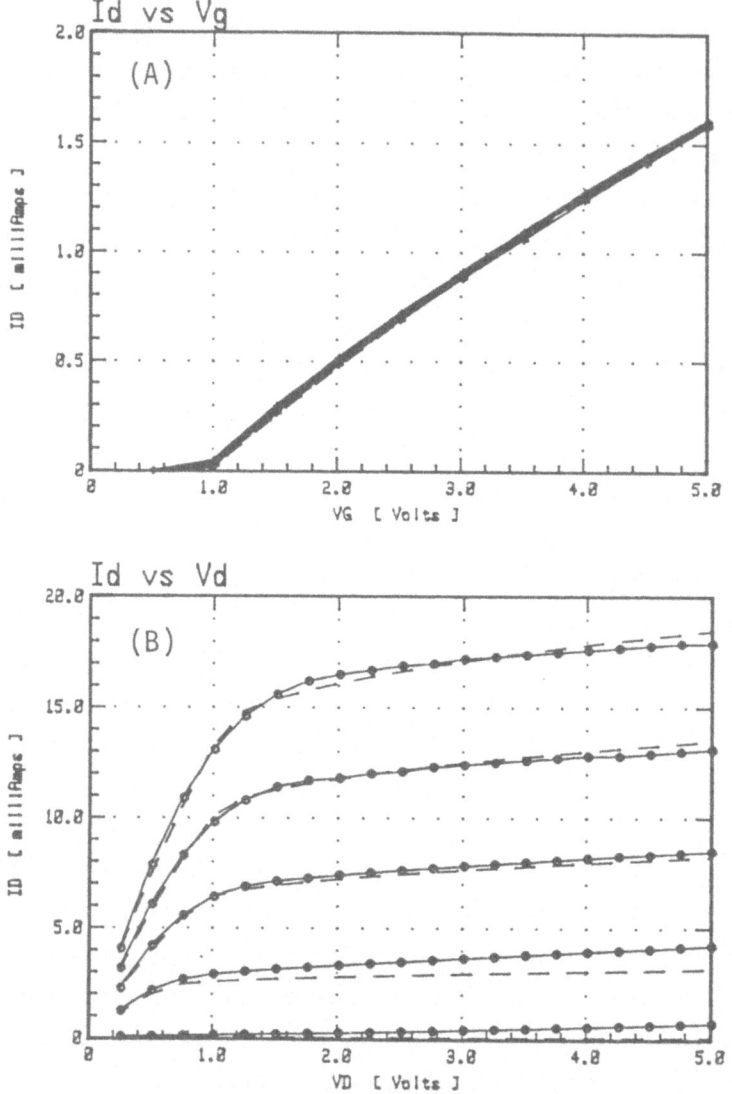

Fig. 12.5 Simulations of the scaled enhancement mode
device. T_{ox} = 25 nm and L_{eff} = 0.75 µm.
The solid lines are by CADDET and the
dashed lines are by SPICE. (A): V_B = -2 to
-5 V; (B): V_G = 1 to 5 V.

A depletion mode device on the same wafer on which the enhancement mode device is fabricated is chosen as a standard device. The mask dimensions of the gate are 3 μm wide and 50 μm long. The structural data of the device are the same as those of the enhancement mode device. The electrical characteristics of the device are shown in Fig. 12.6. The measured data are corrected for the voltage drops by the source and drain resistances. The difference in g_m between $V_{GS} < 0$ V and $V_{GS} > 0$ V indicates the device has a buried channel.

To get the doping profile in the substrate, SUPREM is first run. The profile is converted to a sum of two normal distributions, one for the channel implantation and another for the depletion implantation. Then a CADDET input file is generated to simulate the device. The flat-band voltage (V_{fb}) is adjusted to give the correct magnitude of the drain current at $V_{GS} = 0$ V. Once this adjustment is made, then the simulation of I-V characteristics shows good agreement with measurement in Fig. 12.6.

The scaling of a depletion mode device depends on the scaling result of an enhancement mode device, since the former is used as a load to the latter. In the following, the scaling scheme of a depletion mode device is discussed following the same steps as an enhancement mode device; the reduction in T_{ox}, followed by the reduction in L_{eff}. In each section, it is pointed out how the scaling scheme for a depletion mode device is different from the scheme for an enhancement mode device.

Scaling of Gate Oxide Thickness

The current in the buried channel device in the linear region is closely proportional to the amount of the carrier charge in the channel, which is approximately given by

$$qQ_n \simeq q(Q_{ND} - Q_{NA}) + C_{ox}(V_{GS} - V_{fb})$$
$$\text{when} \quad V_{GS} \geq V_{fb} \tag{12.1}$$

Q_n in the equation is the electron density per unit area. Q_{ND} and Q_{NA} are the dosages of depletion and channel implantations per unit area.

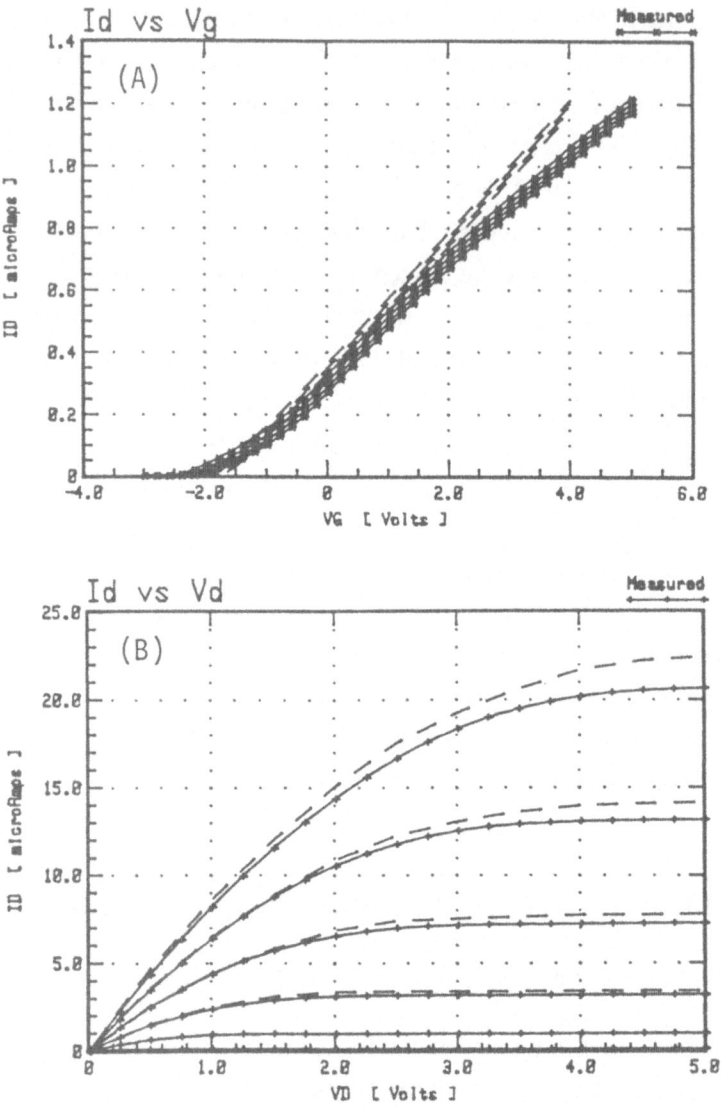

Fig. 12.6 Comparison between the measurements and
the CADDET simulations of a depletion
mode device. $T_{ox} = 40$ nm, $W = 3$ μm and
$L = 50$ μm.   The   solid   lines   are
measurements   and   the   dashed   lines
simulations. (A): $V_B = $ -2 to -5 V; (B): $V_G = $ -2
to 3 V.

The second term in the equation represents the accumulation charge when V_{GS} is greater than V_{fb}. The equation is valid when the depletion boundary lies deeper than the depth of the implantations and $Q_{ND} \gg Q_{NA}$. Eq. (12.1) indicates that the charge in the channel is determined by the net implantation dosage and is independent of C_{ox} when $V_{GS} = V_{fb}$. In case of n$^+$ poly gate device, a flat-band voltage (V_{fb}) is determined by

$$V_{fb} = -\frac{kT}{q}\ln(\frac{N^+}{N_S}) = -0.10 \sim -0.15 \quad V \qquad (12.2)$$

when the voltage drop in the oxide due to the surface state charge (Q_{ss}) is negligible. N^+ in the equation is the doping concentration in the poly gate and N_S is the surface concentration in the substrate. Hence the magnitude of the V_{fb} of an n$^+$ poly gate depletion mode device is close to zero volt.

In real digital circuits, depletion mode devices are used almost exclusively with zero volt from gate to source in the technology being scaled. Hence the device operates practically at flat-band condition. Then, the parameters of importance are the current characteristics at zero gate to source voltage, which are governed by the dosages of implantations, and its sensitivity to substrate bias. Eq. (12.1) can be rearranged to

$$qQ_n \simeq C_{ox}(V_{GS} - V_T) \quad \text{when} \quad V_{GS} > V_{fb} \qquad (12.3)$$

$$\text{where} \quad V_T = V_{fb} - \frac{q(Q_{ND} - Q_{NA})}{C_{ox}} \qquad (12.4)$$

The equation shows that a depletion mode device may be considered like an enhancement mode device when $V_{GS} > V_{fb}$ if we define an extrapolated threshold voltage given by Eq. (12.4). In order to scale the drain current (controlled by Q_{ND} and Q_{NA}) by the same T_{ox} scaling factor as the enhancement mode device, the extrapolated threshold voltage should remain fixed throughout the scaling procedure. In this case, the gate oxide thickness of the depletion mode device is reduced to 25 nm as a result of an enhancement mode device scaling. Also the channel implantation dosage and the energy are set to $1.17E12/cm^2$ and 80 KeV respectively to optimize the enhancement mode devices.

CADDET simulations are performed to study the extrapolated threshold shift by Q_{ND}. The final value of 2.74E12 /cm^2 at 145 KeV is determined for the reduced gate oxide thickness. With the reduced T_{ox} and readjusted Q_{ND}, CADDET is run to get the electrical characteristics as shown in Fig. 12.7. The current path in a depletion mode device is through a buried channel in which the vertical electric field is virtually zero. The mobility degradation by the vertical field, which was a prominent degradation factor in the scaled enhancement mode device does not show up in the scaled depletion mode device. Thus the drain current increases linearly with the scaling factor.

Scaling of Channel Length

The size of a depletion mode device working as a pull-up transistor in an invertor is determined by the size of an enhancement mode device in the circuit and by the specifications for the logic levels of the digital circuit. Since the logic levels and the enhancement mode device are not specified in our example, the mask channel length is varied from 50 μm to 3.5 μm. The drain currents are simulated by CADDET and the results are shown in Fig. 12.8. If we consider the variation of the saturation current by the channel length, it shows that the current gain of a depletion mode device scaled by the channel length is greater than its scaling factor. For example, the saturation current (I_{DS} at $V_{GS} = 0V$, $V_{DS} = 5$ V) increases from 5.93 μA to 136 μA, a factor of 22.9, when L_{eff} is reduced from 49.25 μm to 2.75 μm, a factor of 17.9. The same trend is verified in the experimental measurement. For the case of $L_{eff} = 2.75$ μm, the curve of $L_{eff} = 49.25$ μm is multiplied by the scaling factor (= 49.25/2.75) and plotted in Fig. 12.8. It shows that the current in the linear region scales linearly, while the current in the saturation region

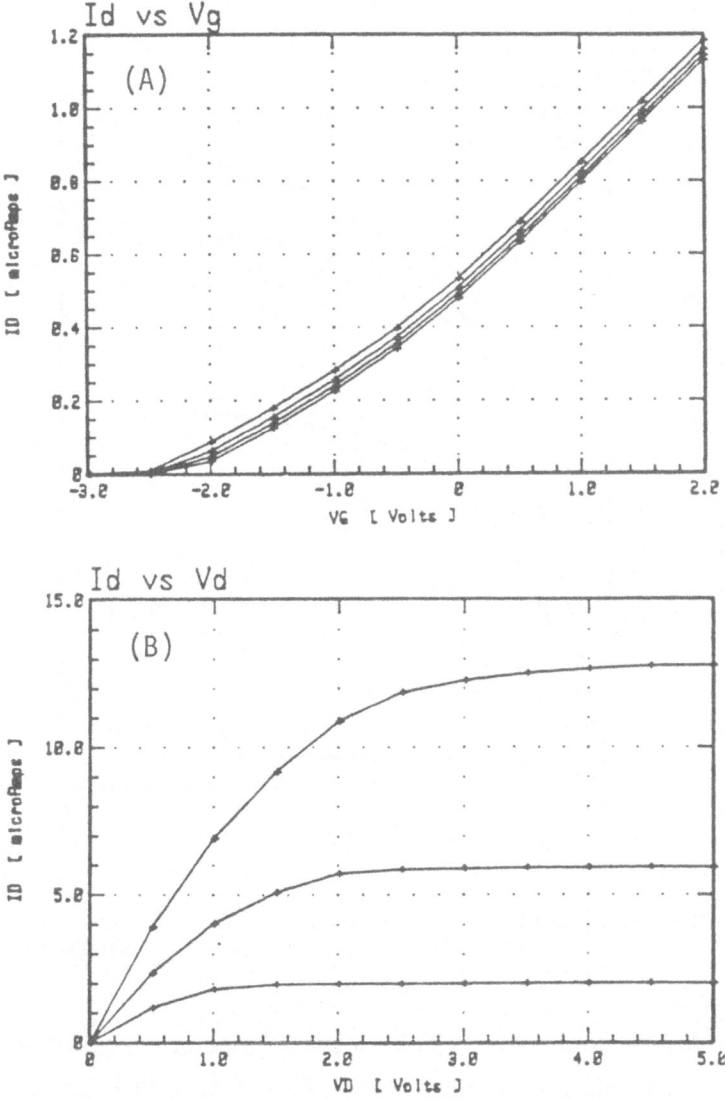

Fig. 12.7 CADDET simulations of the scaled depletion
mode   device   with $T_{ox} = 25$ μm   and
$L = 50$ μm. (A): $V_B = -2$ to -5 V; (B): $V_G = -1$
to 1 V.

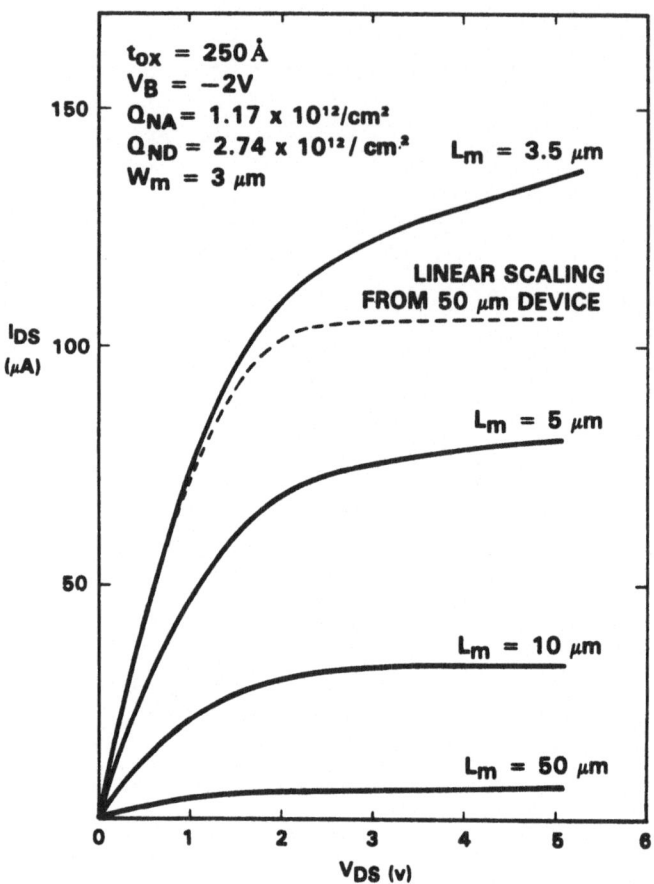

Fig. 12.8 Change of current characteristics of depletion mode devices due to change in channel length, with $V_{GS} = 0$ V.

has high output conductivity. The excessive slope in the saturation region adds extra current gain at $V_{DS} = 5$ V. This extra gain in the current by the L_{eff} scaling is opposite to the case of an enhancement mode device - an enhancement mode short channel device shows mobility degradation already in the linear region and considerable velocity saturation in the saturation region. All these effects combined cause the current gain factor of an enhancement mode device to be smaller than the L_{eff}

W/L [μm/μm]	3/50	3/3.5
T_{ox} [A]	400	250
$\mu 0$ [cm^2/V/Sec]	678	656
V_{TO} [V]	-1.856	-1.825
N_B [/cm^3]	3.16E15	2.30E15
R_S [Ω]	328	328
R_D [Ω]	328	328
ΔL [μm]	0.375	0.375
ΔW [μm]	0.625	0.625

Table 12.3 SPICE parameters for unscaled and scaled
depletion mode devices.

a. Enhancement Mode Devices

T_{ox} [A]	400	250	250
L [μm]	2.0	2.0	1.5
C_{ox} [F/cm^2]	86.3E-9	138E-9	138E-9
V_{TO} [V]	0.674	0.619	0.780
Q_{NA} [/cm^2]	6.0E11	1.4E12	1.17E12
Energy [KeV]	50	50	80
I_{DS}(@ V_{GS}=V_{DS} = 5 V) [A]	9.2E-3	13E-3	18E-3
Current Gain	1.0	1.41	1.96

b. Depletion Mode Devices

T_{ox} [A]	400	250	250
L [μm]	5.0	5.0	3.5
V_{TO} [V]	-1.86	-1.83	-1.83
Q_{ND} [/cm^2]	1.8E12	3.1E12	2.74E12
I_{DS} (@ V_{GS}=0, V_{DS}=5V) [μA]	48.7	80.4	136
Current Gain	1.0	1.62	2.79

Table 12.4 Summary of device scaling.

scaling factor. The device parameters of the unscaled and the scaled depletion devices are shown in Table 12.3.

12.4 Conclusions

It is shown that the MOSFET devices can be scaled to improve device performance. In case of an enhancement mode device, gate oxide thickness and channel length are chosen as prime scaling factors. The reduction in the gate oxide thickness increases the drain current, but the surface mobility degrades and this causes deviation from the linear scaling. The reduction in channel length promotes short-channel effects, especially the saturation of the carrier velocity. After the punchthrough voltage is optimized, the final gain in the current is 1.96.

In case of a depletion mode device, the depletion implantation dosage and the channel length become the major scaling factors. Since the current path is through the buried channel, the mobility is insensitive to the vertical field from the gate terminal. However, the excessive output conductance in the saturation region of the reduced channel length should be carefully controlled.

Table 12.4 shows the summary of overall MOSFET scaling. CADDET is used intensively along with SUPREM. Their usefulness in studying the scaling schemes is well demonstrated.

References

[12.1] R. H. Dennard et al, "Design of Ion-Implanted MOSFET's with Very Small Physical Dimensions," *IEEE J. Solid-State Circuits*, <u>SC-9</u>, No.5, 1974, pp 256.

[12.2] E. H. Nicollian and J. R. Brews, *MOS Physics and Technology*, New York: John Wiley & Sons, Inc., 1982.

[12.3] R. C. Y. Fang, R. D. Rung and K. M. Cham, "An Improved Automated Test System for VLSI Parametric Testing," *IEEE Trans. Instru. Meas.*, IM-31, no. 4, Sept 1982, pp. 198-205.

[12.4] L. A. Akers and J. J. Sanchez, "Threshold Voltage Models of Short, Narrow and Small Geometry MOSFET's: A Review", *Solid-State Electronics*, 25, No.7, pp 621-641, July 1982.

[12.5] D. A. Antoniadis, S. E. Hansen, and R. W. Dutton, "SUPREM II - A Program for IC Process Modeling and Simulation," TR 5019.2, Stanford Electronics Laboratories, Stanford University, Calif., June 1978.

[12.6] A. S. Grove, *Physics & Technology of Semiconductor Devices*, New York: John Wiley & Sons, Inc., 1967.

[12.7] L. D. Yau, " A Simple Theory to Predict the Threshold Voltage of Short-Channel IGFET's", *Solid -State Electronics*, 17, pp. 1059-1063, Oct 1974

Chapter 13

Parasitics Extraction for VLSI Process Development

13.1 Introduction

Circuit performance in general depends on two major factors. These are transistor performance, and parasitic capacitances and resistances. The scaling down of geometrical dimensions introduces two-dimensional and even three-dimensional effects both in transistor behavior and in the parasitics. The scaling of interconnections is usually emphasized primarily in the width and spacing of the lines. For example, the line/space design rule (in micrometer units) for the first level aluminum in CMOS SRAMs has been scaled from 2.5/3.5 in 16Kb chips to 2/2 in 64Kb chips to 1.2/1.6 in 256Kb chips [13.1][13.2]. The line thickness is often scaled down by a smaller factor in order to reduce the parasitic resistance, and to increase the reliability against electromigration [13.3][13.4]. The increase in aspect ratio of the interconnect line thickness to line width and space increases the fringing and interline capacitance component in the total parasitic capacitance of a circuit. This means that the parasitic capacitance is not scaled down proportionally as the horizontal dimensions are scaled down.

In VLSI circuits, with the number of transistors approaching one million, the interconnections become a major issue in the circuit performance. Also as the transistor channel length is scaled down to the

283

range of one micron or below, velocity saturation will limit the gain in the current drive of the transistors [13.5]. Hence transistor performance will improve only slowly with scaling. Many creative techniques are implemented to reduce the resistance and capacitance of the interconnections. In order to be able to predict the speed of VLSI circuits, as well as to understand the relationship between parasitic capacitance and interconnect structures and layout, accurate extraction of parasitics is of great importance in VLSI. Also, if experimental measurements of the parasitics (which are often indirect measurements, meaning that two or more measurements are made on two or more structures, then are followed by calculations based on simple assumptions) can be verified by simulations, the parameters extracted will have a

PROCEDURE FOR PARASITIC CAPACITANCE EXTRACTION

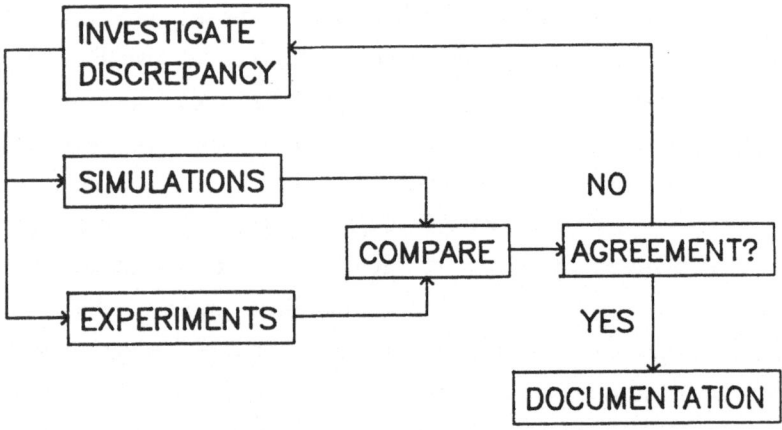

Fig. 13.1 Methodology in parasitics extraction.

higher confidence level. The simulations provide more physical insight into the structures and their contributions to the parasitics, since they provide information about the internal electric potential profiles.

Fig. 13.1 shows the methodology in the extraction of parasitic capacitance. The experimentally extracted values are compared with simulations. Any inconsistency must be investigated until it is understood.

Simulations can provide an estimate of the parasitics if no test structure is available for measurement, or if that direct measurement cannot be easily performed, such as the case of capacitive coupling between three or more layers. It can provide guidance in the development of new processes, and allow one to see if there is any significant performance improvement before the process is actually developed. It can also provide guidance in process development such as choosing the appropriate dielectric between the interconnect layers, since the dielectric constant will affect the parasitic capacitance.

13.2 Simulation Techniques

The simulation system for parasitic capacitance extraction is shown in Fig. (13.2). The system provides the capability of two-dimensional simulations of interconnect structures. Multilayer and interline coupling can also be studied.

Parasitic Simulation by FCAP2

The areal and peripheral coefficients of parasitic capacitances between conductive layers can be investigated using the FCAP2 program. The structures range from the simple case of conducting lines over field oxide to more complicated cases of multilayer coupling and interline coupling. Examples are:

    (1) metal lines over field oxide, polysilicon layers, or diffusion

    (2) polysilicon (poly) lines over field oxide

    (3) poly lines over field oxide covered by metal plane

    (4) interline coupling of metal lines over poly plane

Since FCAP2 does not accept semiconductor layers, the silicon layers are assumed to be conductors. This is a reasonable assumption for

2D SIMULATION SYSTEM FOR PARASITIC CAPACITANCE EXTRACTION

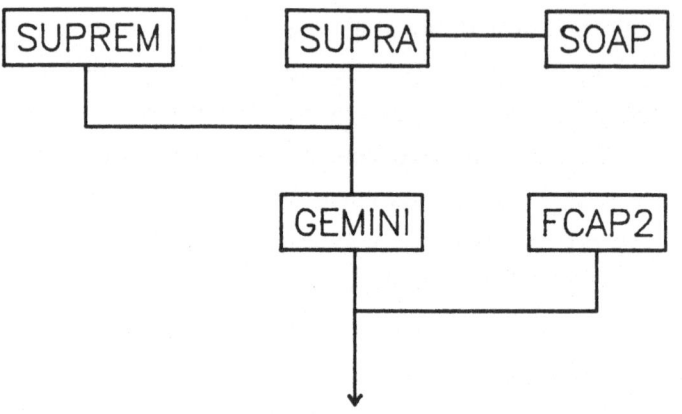

PARASITIC CAPACITANCE COEFFICIENTS

Fig. 13.2 Simulation system for parasitic extraction.

case (1), where the dielectric thickness is much larger than the depletion layer thickness in the silicon substrate. However, when the polysilicon line goes over the field oxide (case (2) and (3)), it forms a depletion layer on the surface of the substrate, the width of which is dependent on the impurity profile under the field oxide, and can be a significant fraction of the field oxide thickness. The GEMINI program is used for this case. This technique will be discussed in the next section.

The simplest calculation would be that of a conducting line over a conducting plane (such as diffusion, or silicon substrate) separated by a layer of dielectric such as silicon dioxide, representing case (1). Fig. 13.3 shows such a simulation. The equipotential lines are shown (the electric field lines are perpendicular to the potential contours), which clearly shows the importance of the fringing component, i.e., the component of capacitance arising from the two sides of the conducting line. For VLSI, the dielectric stru ture can be more complicated, and may contain a

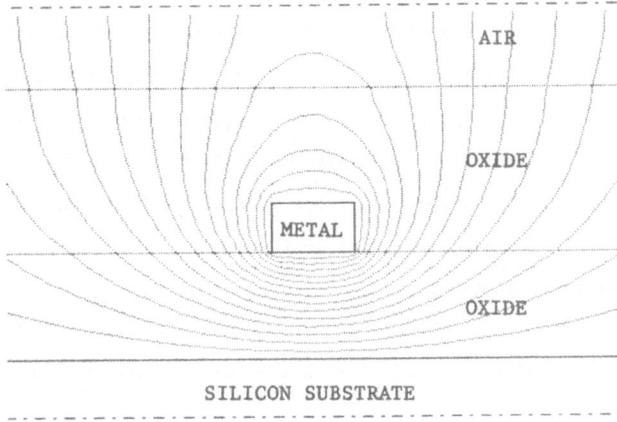

Fig. 13.3 FCAP2 simulation of a metal line over field
oxide.

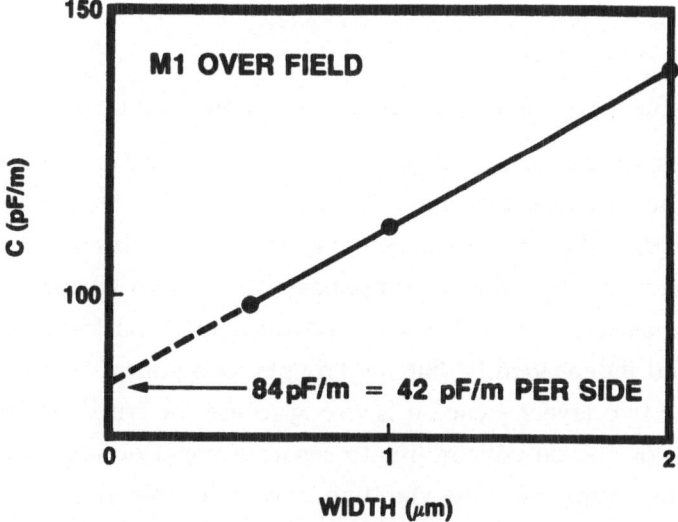

Fig. 13.4 Extraction of the fringing capacitance of a
metal line over field oxide.

combination of different dielectric materials. This is due to issues such as planarization for better line width control. FCAP2 can also simulate these cases. The areal and fringing components of the capacitance are extracted by performing the simulation with different widths of conducting line. By plotting the total capacitance versus the line width, the slope of the resulting line and the y-intercept will provide the areal and fringing components of the parasitic capacitance, respectively. Fig. 13.4 shows the result for the case where the interlevel dielectric is a combination of nitride and oxide.

As an example of the study of a more complicated structure, FCAP2 is used to simulate case (3) although the approximation that the substrate is

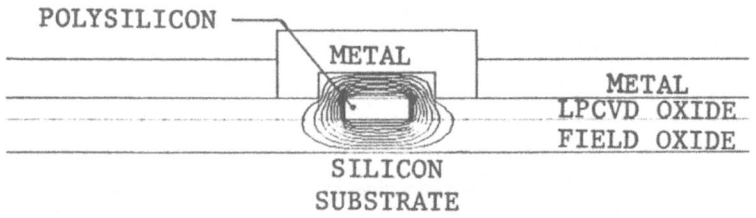

Fig. 13.5 Multilayer capacitive coupling simulation.

metallic has to be made. Fig. 13.5 shows the structure generated by FCAP2. The program calculates the charge induced on the metal and substrate, hence the capacitive coupling to the two layers. The major interest here is in the fringing components to the two layers. The areal capacitance between the polysilicon and the upper and lower layers can be calculated independently, but the fringing component will be "shared" between the two layers. Hence it is wrong to use the fringing capacitance coefficients of the polysilicon line to separate metal or substrate planes. The fringing component to the two layers in this structure can be simulated by using essentially the same method as described in the previous discussion of a conducting line over a conducting substrate. By varying the polysilicon line width, the fringing capacitance to the metal

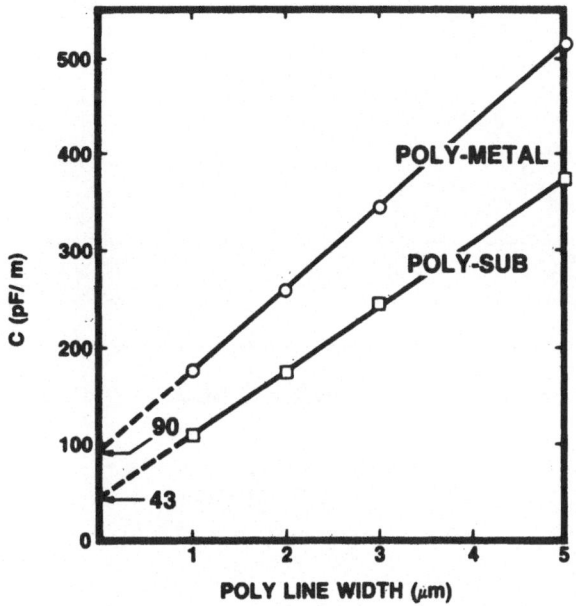

Fig. 13.6 Extraction of multilayer fringing capacitances.

and substrate can be calculated. Fig. 13.6 shows the extraction of the parameters. The simulation shows that when the poly line is covered by metal on the top, the fringing capacitance of the poly line to the substrate is reduced by 60%, since part of the fringing field has terminated at the metal plane.

Another interesting situation is for the case of metal over diffusion. Fig. 13.7 shows the two cases of metal line over diffusion plane (upper figure) and metal plane over diffusion line. The measured parasitic capacitance (in units of $fF/\mu m^2$ for areal capacitance CA and $fF/\mu m$ for fringing capacitance CP) are shown for both cases. The thickness of the metal, low temperature oxide (LTO) and field oxide (F.OX) are 1.0, 0.5 and 0.5 microns respectively. The fringing capacitance for the diffusion line to metal plane is much smaller than that of the metal line to diffusion plane, and cannot be explained by the difference in the

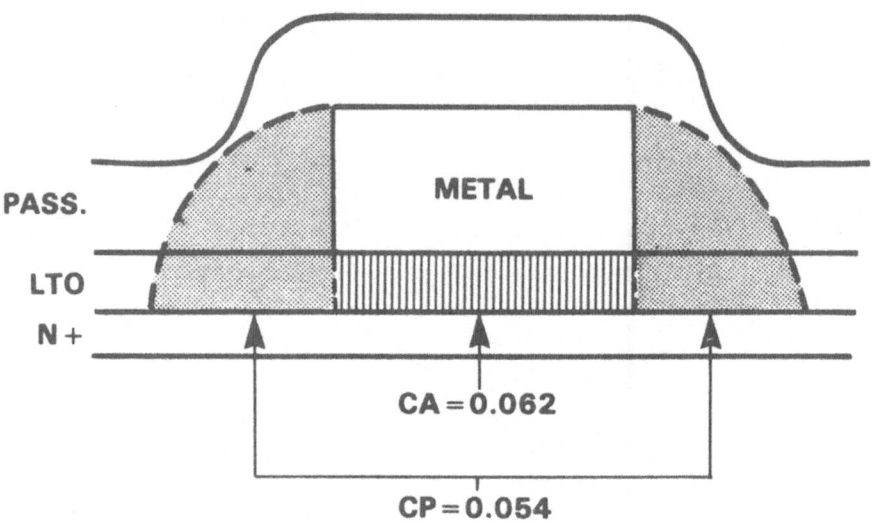

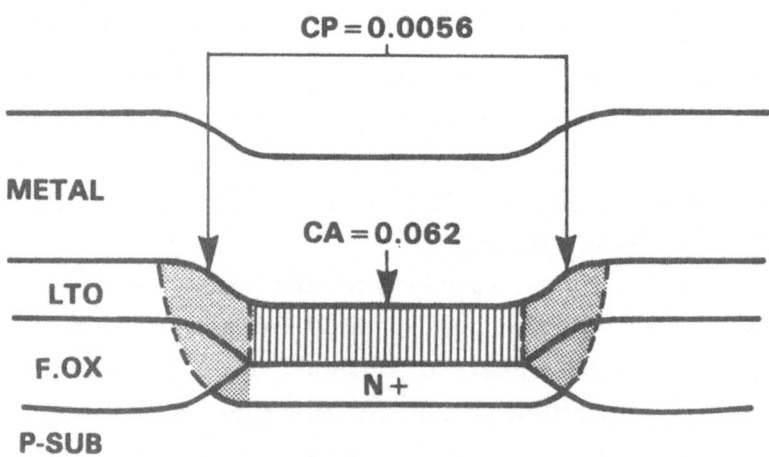

Fig. 13.7 Comparison of fringing capacitance between
metal line to n⁺ plane and n⁺ line to metal
plane; units are in fF/μm² and fF/μm for
area and peripheral components respectively.

thickness of the two layers alone. This is because at the side wall of the diffusion line, the effective dielectric thickness between the metal plane and the diffusion perimeter is much larger than the deposited oxide thickness. This is due to the shape of the field oxide in that region. The ability to accept arbitrary shapes for the Laplace equation makes this structure acceptable for FCAP2 simulations.

As mentioned in the introduction, the interline capacitance of interconnections is becoming increasingly significant as the line spacings are reduced while the line thickness remains about the same. FCAP2 is used to simulate the case of metal lines with small spacings, and over a polysilicon plane. The simulated potential contour for the case of metal lines having different potentials is shown in Fig. 13.8. In this case, the line spacing is 0.8 μm and the line thickness is 0.6 μm. Fig. 13.9 shows the graph of the capacitance of the metal line at 3 V versus line spacing for a case where the interlevel dielectric is a combination of nitride and oxide. The results show that the fringing capacitance increases by 3E-3 fF/μm and 1.3E-2 fF/μm when the line spacings are 2 μm and 1 μm respectively. The increase in capacitance is also dependent on the distance of the lines from the underlying conducting plane. If the distance is reduced, the interline capacitance will be reduced, since more of the field lines will be terminated on the conducting plane, thus reducing the coupling between the metal lines. Fig. 13.10 compares the two cases where the metal lines are above the substrate and above the polysilicon plane. The simulated results show that the increase in the total metal line capacitance with reducing line spacing (ΔC) is larger for the former case.

In the case where the potentials of the metal lines are the same, there will be a reduction of the fringing capacitance due to electric field shielding. (The extreme case would be when the line spacing is zero, in which case the total fringing capacitance due to the side walls of the two lines would be reduced by a factor of two.) Fig. 13.11 shows the potential contour for this case. By using the same technique as was discussed for the case of lines with different potentials, the reduction in

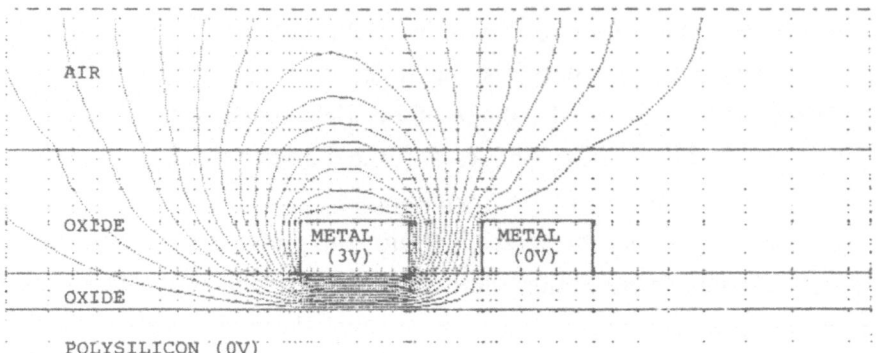

Fig.13.8 2-D potential profile for two metal lines at
        different potentials, above polysilicon plane.

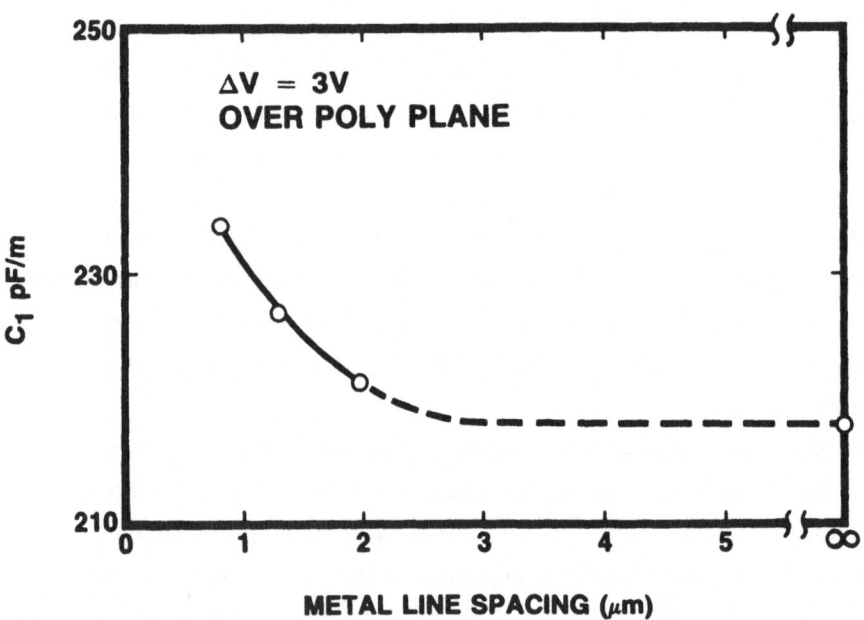

Fig. 13.9 Simulated capacitance per line vs. spacing for
         two metal lines at different potentials.

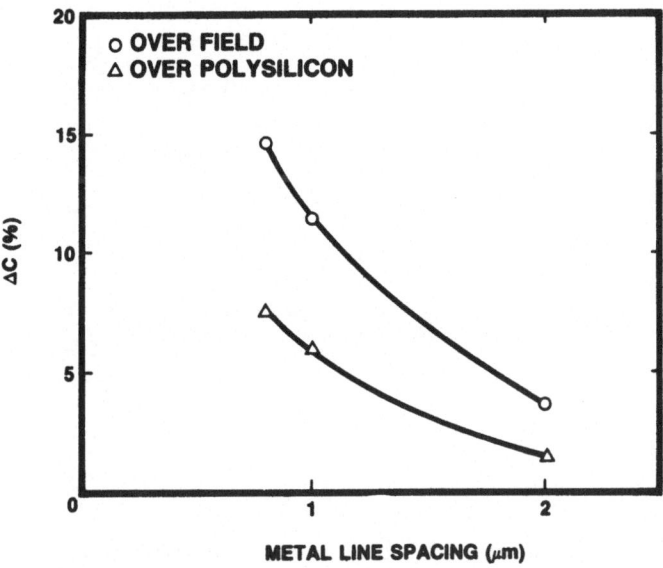

Fig. 13.10 Simulated increases in metal line capacitance vs. metal line spacing for two metal lines at different potentials.

the parasitic capacitance is calculated, as shown in Fig. 13.12.

The interline capacitance, for a certain line spacing, is equal to the charge induced on the metal line per volt of change in the potential difference between the lines. This capacitance can be found by calculating the difference in the capacitance between the graphs shown in Fig. 13.9 and 13.12. For example, the capacitive coupling between the metal lines at a spacing of 1 μm is 43 pF/m, which about 20% of the capacitance of a single metal line.

Field Capacitance by SUPREM and GEMINI Simulations

When the capacitance involves the formation of a depletion layer with a width which is a significant fraction of the dielectric thickness, the FCAP2 program is inaccurate. This occurs in the case of the field capacitor, where the substrate underneath the field oxide forms a depletion layer, the depth of which depends on the field implant profile.

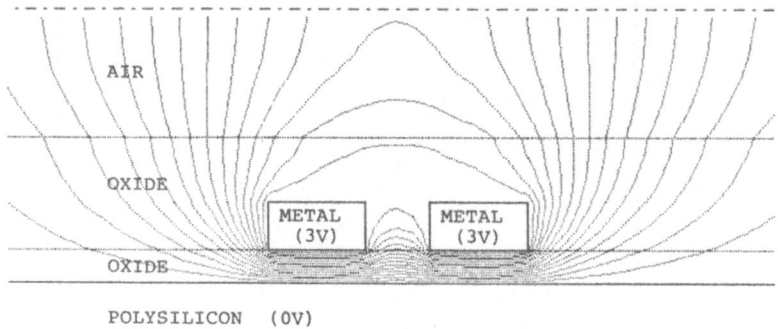

Fig. 13.11 Simulated 2-D potential for two metal lines
at the same potential, over polysilicon plane.

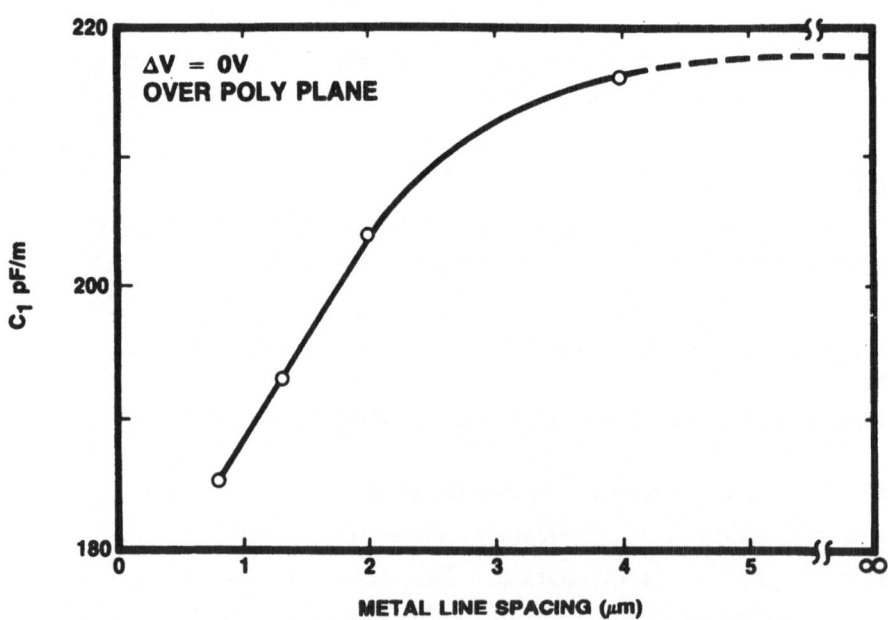

Fig. 13.12 Simulated capacitance per line vs. metal line
spacing for two metal lines biased at the
same potential, over polysilicon plane.

In this case, the depletion layer thickness is about 0.1 μm, while field oxide is typically 0.5 to 0.6 μm. Poisson's equation has to be solved for a given bias condition, and the capacitance calculated.

The SUPREM program is used to simulate the impurity profile of the field implant under the field oxide. Then the profile is transferred to the GEMINI program, which creates the field capacitor, with the desired bias

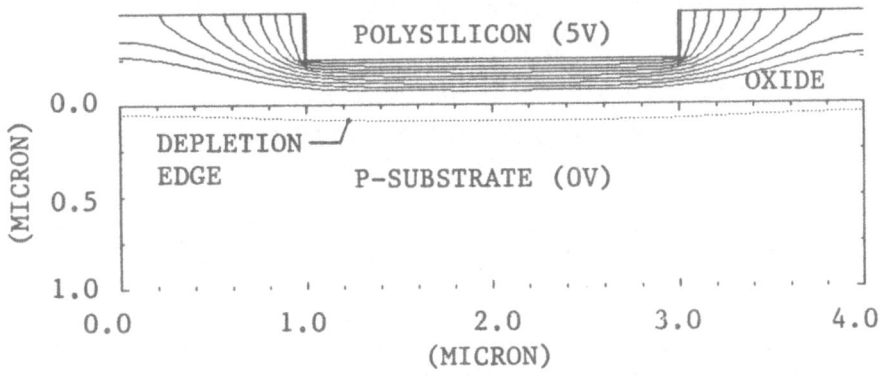

Fig. 13.13 GEMINI simulation of poly field capacitor.

on the capacitor electrode and substrate. Fig. 13.13 shows the potential contours of a GEMINI simulation of a polysilicon field capacitor. The symmetric boundary condition for the potential also allows one to consider the effect of periodic poly lines biased at the same potential and separated by a specified spacing. This condition occurs in the poly serpentine structure, and has the effect of reducing the fringing capacitance. Fig. 13.14 shows the extraction of the fringing and areal capacitances for an essentially isolated polysilicon line. The depletion layer has significantly reduced the parasitic capacitance. Fig. 13.15 shows the effect of varying spacing between poly lines at the same potential, with fixed line width. The reduction in the total capacitance is due to the reduction of the fringing component. The results shows that the reduction is not significant until the spacing is below 2 μm.

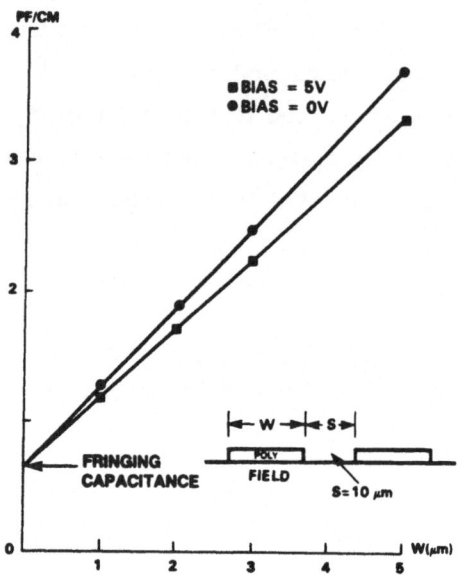

Fig. 13.14 Polysilicon line to field areal and fringing
           capacitance extraction.

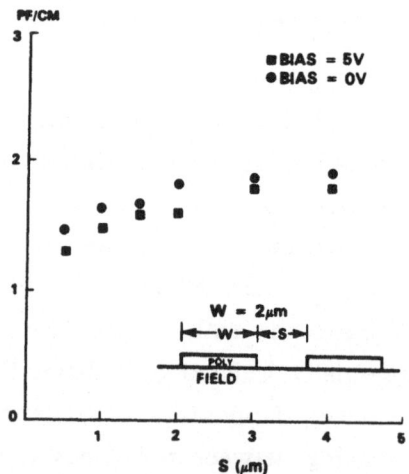

Fig. 13.15 Simulated   polysilicon   line   over   field
          fringing capacitance vs. line spacing, for
          lines at the same potential.

13.3 Diffusion Capacitance by SUPRA Simulations

Diffusion capacitance simulation involves the simulation of the electric field in heavily doped regions. The capacitance is a function of the impurity profile in the structure. Fig. 13.16 shows the simulation by SUPRA of the impurity profile of an n^+ diffusion region adjacent to the

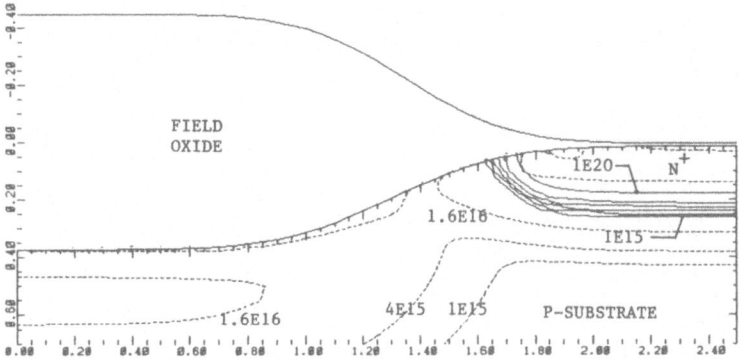

Fig. 13.16 2-D simulation of the n^+ profile by SUPRA. The dotted and solid lines are for boron and arsenic concentration (cm^{-3}) contours respectively.

field oxide. The accuracy of the capacitance calculations depends very much on the accuracy of the n^+ profile. Results have shown that more accurate data are obtained by measurements compared with simulations at the present time. Thus measured data are used for circuit simulations. It is still useful to simulate the profile in this case because it provides a general picture of the impurity profile, and also indicates what will happen qualitatively to the capacitance if the processing steps are changed. The errors of the capacitance calculation reflect the difficulty of accurate process simulation for impurity profiles, particularly at the edges.

13.4 Parasitic Resistance by FCAP2 Simulations

The resistance of a film of arbitrary two-dimensional geometry can be calculated using the FCAP2 program. Fig. 13.17 shows the simulation of

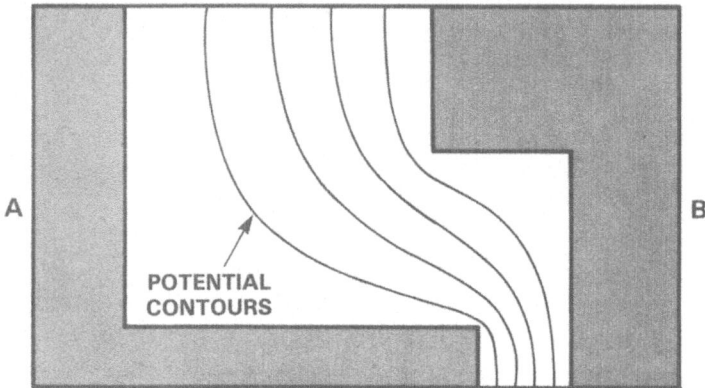

Fig. 13.17 Resistance calculation for arbitrary 2-D shape.

the resistance of a two-dimensional shape. In order to use FCAP2 to calculate the resistance from contact (shaded areas) A to B, the materials of the contacts are defined to be metallic, and set at different potentials. The capacitance between the two metallic contacts are then calculated by FCAP2. The capacitance is related to the surface integral of the electric field at the metallic surface (the contacts A and B). This result can be converted into current flowing between A and B by replacing the dielectric constant between the contacts used in the capacitance calculation by the conductivity of the material under study. The total current flowing through the two contacts can thus be found for the bias condition specified, hence giving the resistance between the contacts. In this example, the calculated effective sheet resistance is 0.38*(sheet resistivity). In other words, the 2-D shape between the two contacts is equivalent to 0.38 square in resistance.

13.5 Experiment and Simulation Comparisons

Measurements have been made on parasitic capacitance test structures and are compared with simulated results. Examples of comparisons are shown in Table (13.1). The physical dimensions of the test structures were measured by SEM, and used in the simulations. Results show that

	C(AREA) FF/UM**2	C(PER) FF/UM
POLYSILICON LINE OVER FIELD DATA: SIMULATION:	0.053 0.056	0.046 0.035
METAL LINE TO DIFFUSION DATA: SIMULATION:	0.062 0.069	0.061 0.060
METAL LINE TO POLYSILICON DATA: SIMULATION:	0.067 0.069	0.061 0.058

Table 13.1 Data-simulation comparisons of parasitics.

the agreement is, in general, quite good. Note that in several cases of multilayer structures such as poly lines over field oxide and under a metal plane, a combination of simulations and measurements is necessary to obtain the coefficients. Much improvement in the accuracy of circuit simulations has been obtained by using this methodology of parasitic capacitance extraction.

13.6 Summary

Accurate simulations of parasitic capacitances are found to be very essential in VLSI process development. Parasitics have become the major factor in speed performance, and structures contributing to parasitic capacitances are getting more complicated due to the higher complexity of interconnect layers. It has been found that simulations can provide insight into the problem, provide guidance in process development, as well as higher confidence in the extracted parameter values. The interconnections are becoming more complex as more conducting levels are added. The number of conductor cross-overs has become so large in VLSI that both node-to-node coupling and total capacitance are affected. Three-dimensional calculations are necessary to accurately predict the resulting circuit effects. Some progress has been made by researchers [13.6].

References

[13.1] S. Konishi, et al, "A 64Kb CMOS RAM," *Tech. Digest of ISSCC 1982*, pp. 258-259.

[13.2] M. Isobe et al, "A 46ns 256K CMOS RAM," *Tech Digest of ISSCC 1984*, pp. 214-215.

[13.3] J. Black, "Physics of Electromigration," *Proc. 12th Reliability Physics Symposium*, IEEE, New York, 1974, pp.142.

[13.4] S. Vaidya, D. B. Fraser, and A. K. Sinha, "Electromigration Resistance of Fine Line Al," *Proc. 18th Reliability Physics Symposium*, IEEE, New York, 1980, pp. 165.

[13.5] Y. El-Mansy, "MOS Device and Technology Constraints in VLSI," *IEEE Trans. in Electron Devices*, ED-29, Apr 1982, pp. 567-573.

[13.6] M. Fukuma and R. H. Uebbing, "Wiring Capacitance Simulation in Two and Three Dimensions," *Tech. Digest of Symposium on VLSI Technology 1984*, pp. 24-25.

Appendix

Source Information of 2-D Programs

SOURCE INFORMATION OF 2-D PROGRAMS

PROGRAM	SOURCE	ADDRESS	
SUPREM	Stanford Univ.	Office of tech. licensing, Stanford Univ. 105 Encina Hall, Stanford, CA 94305	Public
SUPRA	•	•	•
SOAP	•	•	•
GEMINI	•	•	•
CADDET	Hitachi	Tech. Administration, Hitachi P.O. Box 2, Kokubunji, Tokyo, Japan	Technical exchange
SIFCOD	Michael Sever (Mock)	24 Dubnov St., Tel Aviv 64332 Israel	Commercial
FCAP2	Hewlett Packard	Hewlett Packrad Lab. 3500 Deer Creek Rd., Palo Alto, CA 94304	Technical exchange
TECAP2	•	•	Commercial
HPSPICE	•	•	Commercial

Appendix
Some Information on PCBs

Table of Symbols:

A	angstrom unit (1E-10 m)
B	parabolic oxide growth-rate constant
B/A	linear oxide growth-rate constant
BV_{DS}	source-drain breakdown voltage
C	capacitance
C^*	equilibrium oxidant concentration in the oxide
C^o	oxidant concentration at the oxide interface
C_1, C_2	doping concentrations
C_A	areal capacitance coefficient
C_D	depletion capacitance
C^i	oxidant concentration at the silicon interface
C_n	coefficient of auger recombination
C_{ox}	gate oxide capacitance
C_P	peripheral capacitance coefficient
C_p	coefficient of auger recombination
C_p	peak concentration
C_T	total impurity concentration
D_C	counter-doping dose
D_{eff}	effective diffusion coefficient
$D_i{}^+$	diffusivity due to positive vacancies
$D_i{}^-$	diffusivity due to negative vacancies
$D_i{}^=$	diffusivity due to doubly negative vacancies

$D_i{}^x$	diffusivity due to neutral vacancies
d_k	mask thickness at k^{th} segment
D_N	diffusivity under non-oxidizing condition
E	electric field strength
E_c	electron energy of conduction band edge
E_{fp}	quasi-Fermi level for holes
E_i	electron energy at the intrinsic Fermi level
E_m	maximum electric field
E_n	electric field normal to current flow
E_p	electric field parallel to current flow
E_{SAT}	critical field for velocity saturation
E_μ	activation energy of oxide viscosity
F, fF, pF	Farad unit, 1E-15 F, 1E-12 F
F, F_1, F_2, F_3	oxidants fluxes
G	mobility coefficient
G_{DS}	channel conductance
g_m	transconductance
g_z	oxide growth rate in the z-direction
h	gas transport coefficient
I	current
I_{DD}	maximum available current from power supply
I_{DS}	drain to source current
I_{DSAT}	drain to source current at saturation
I_H	latchup holding current
I_L	transistor subthreshold leakage current
I_{max}	maximum concentration
I_{PT}	punchthrough current
I_{SUB}	substrate current
I_T	latchup triggering current

I_{TH}	transistor threshold current
$\mathbf{J}$	impurity flux
$\mathbf{J_n}$	electron current density
$\mathbf{J_p}$	hole current density
k	Boltzmann constant
k	surface reaction coefficient
L	channel length
L	decay length in thin oxide regime
L_D	length of uniform drain region
L_{eff}	effective channel length
L_S	length of uniform source region
L_v	diffusion length of vacancies
m	segregation coefficient
$\mathbf{n}$	unit vector normal to the surface
n	electron density
N	$N_A\text{-}N_D$
N_o	coefficient of band gap narrowing
N_o	concentration redistributed after oxidation
N_{ox}	oxidation concentration in the x-direction
N_{oy}	oxidation concentration in the y-direction
N^+	doping concentration in n+ polysilicon
N_A	acceptor concentration
N_B	substrate doping
N_c	correction concentration for oxidation
N_D	donor concentration
N_i	intrinsic carrier density
n_i	effective intrinsic carrier density
$n_{i,e}$	inert drive-in concentration
N_{ix}	inert drive-in concentration in the x-direction

N_{iy}	inert drive-in concentration in the y-direction
N_{iz}	oxidation concentration in the z-direction
nm	nanometer unit (1E-9 m)
N_p'	implantation dose
N_S	substrate surface impurity concentration
n_S	surface electron concentration
P	pressure
p	hole density
q	electron charge
Q_n	electron density per unit area in the channel
Q_{NA}	channel implant dose
Q_{ND}	depletion implant dose
Q_{ss}	interface fixed charge density
R	recombination
R_D	drain series resistance
R_p'	range of implantation
R_S	source series resistance
S	subthreshold slope
T	absolute temperature
T_{ox}	oxide thickness
U	active impurity concentration
V	oxide growth velocity
V^-	negative vacancies
$V^=$	doubly negative vacancies
V^+	positive vacancies
V_B	substrate bias voltage
V_{BS}	substrate to source bias voltage
V_c	mobility coefficient

V_D	drain bias voltage
V_{DD}	circuit bias voltage
V_{DS}	drain to source bias voltage
V_{DSAT}	saturation drain voltage
V_{fb}	flat-band voltage
V_G	gate bias voltage
V_{GS}	gate to source bias voltage
V_{in}	invertor input voltage
V_{out}	invertor output voltage
V_{PT}	punchthrough voltage
V_s	saturation velocity
V_S	source bias voltage
V_{SS}	circuit ground voltage
V_T	transistor threshold voltage
V_{T0}	threshold voltage at zero substrate bias
V_{TN}	n-channel threshold voltage
V_{TP}	p-channel threshold voltage
V^x	neutral vacancies
W	transistor channel width
W_D	drawn channel width
W_{eff}	effective channel width
x	x-coordinate
X_i	initial oxide thickness
X_j	diffusion junction depth
X_d	depletion depth
y	y-coordinate
Y_j	counter-doping junction depth
ΔL	channel length loss per side
ΔR_p	standard deviation in the y-direction
ΔW	channel width loss per side
∇	gradient operator

ψ	internal potential
ψ_B	difference between intrinsic and hole Fermi level
ϵ	dielectric constant
ϕ_p	hole quasi-Fermi potential
ϕ_n	electron quasi-Fermi potential
Θ	vector stream function
Θ	z-component of stream function
μ	mobility
μ	oxide viscosity
μ_o	zero-field mobility
μ_n	electron mobility
μ_p	hole mobility
ρ	oxide density
σ	standard deviation of implantation
τ_n	electron life time
τ_p	hole life time
$\delta\phi_n$	increment of electron quasi-Fermi potential
$\delta\phi_p$	increment of hole quasi-Fermi potential
2-D	two-dimensional
1-D	one-dimensional

Subject Index

Automatic grid generation, 86, 113

Avalanche breakdown, 135, 195, 240

BIRD, 14

Bird's beak, 201, 223

Body effect, 153-155

Boron encroachment
  LOCOS, 201, 218, 223
  modified LOCOS, 223-227
  SWAMI, 232-234

Boundary-value method, 56

Breakdown, source-drain, 135, 195, 240

Buried channel, 273

CAD, 123

CADDET
  basic equation, 84-85
  device simulations, 242-244
  flow chart, 88
  grid, 86-87
  input file, 93
  input format, 92-94
  input structure, 83

CADDET (cont.)
  numerical technique, 86
  output, 95

Capacitance
  areal, 288
  depletion layer, 133, 294-295
  diffusion, 297
  extraction, 284, 286-288
  fringing, 287-290
  gate oxide, 133
  interline, 283, 291-293
  metal line over field, 286-288
  multilayer, 288-289
  poly over field, 293-296

Channel electric field, 240

Channel implant
  deep, 150
  depletion, 273-276
  n-channel, 150-152
  shallow, 150

Channel length
  effective, 148, 262

Channel profile
  n-channel, 142, 194
  p-channel, 179, 189

Channel width, 218
  drawn, 218
  effective, 148, 218
  effective, extraction of, 222-223
  loss, 222-223, 235
Channeling, 179, 189
CMOS, 171-173
  cross-section, 25
  n-well, 200
  p-well, 213, 215
  submicron, 171-173
  trench-isolated, 199
Computer-aided design, 123
Counter-doping, 124, 179
  dose, 124, 178
  energy, 124
  junction, 179, 180
Critical electric field, 240
Current continuity equation
  stream function form, 85
  time dependent form, 98-99
Current transport equation 98-99

Degradation
  hot electron, 195-196, 240
  linear transconductance, 245
  saturation transconductance, 245
Depletion layer
  capacitance, 133
Depletion mode MOSFET, 271-279
Device optimization, 138
Device simulation
  history, 13-14
DIBL, 159-169

Dielectric relaxation time, 99
Diffusion
  diffusion continuity equation, 30
  field-driven, 31
  gradient-driven, 31
  high concentration, 44-46
  low concentration, 43-44
  moving boundary, 44-46
  vacancy diffusion, 31
Diffusivity
  boron, 31
  effective, 54
  intrinsic, 31
  phosphorus, 32
Double diffused drain, 195, 241
Drain electric field, 248-258
Drain-induced barrier lowering,
         133, 159-169
  p-channel, 179-191

Effective channel length, 148, 262
Effective channel width, 148
Effective intrinsic carrier density,
         100
Einstein's relation, 85
Electric field
  channel, 239
  critical, velocity saturation, 240
  drain, 239-240, 248-258
  maximum, drain, 240, 253-254
  normal, 89
  parallel, 89
Electron mobility, 89, 100, 262
Enhancement mode MOSFET,
         262-271

Factorial experiments, 124
FCAP2
  applications, 115
  boundary conditions, 113
  charge calculation, 117
  contour plot, 117,119
  flow chart, 116
  grid generation, 113
  input file, 118
  input format, 114-115
  parasitic simulations, 285-294
  remap, 119
Fermi-Dirac statistics, 84
FIELDAID, 13
Finite difference method, 73
Flat-band voltage, 275
Fringing capacitance, 283, 287-290
Fully-recessed oxide isolation
  SUPRA simulation, 48-49

Gate oxide
  capacitance, 133
  effect on subthreshold slope, 185
GEMINI
  basic equation, 69
  capability, 66-67
  channel length simulation, 74-78
  channel width simulation, 79-81
  device simulations, 141, 145
  five-point finite-difference
        approximation, 73
  grid, 71
  input file, 75, 80
  numerical techniques, 73

GEMINI (cont.)
  punchthrough current
        calculation, 70-71
  SWAMI simulation, 228-234
  trench simulation, 202-205
Glass-transition temperature, 54
Graphical post processing, 16
Green's function, 58
Gummel's iteration, 99,102,105

Hierarchical simulation, 15
Hot electron degradation, 138,
        195-196, 240
HP-SPICE, 19

ICCG, 113
ICCG3, 97
Impact ionization, 240
Implantation, energy, 270
Impurity profile
  n-channel, 142, 145
  source/drain, 143
  SWAMI, 232
  trench structure, 205
Interface damage, 240
Interline capacitance, 283, 291-293
Invertor circuit, 145, 268
Ion implantation
  analytical model (SUPRA), 41-42
  boron, 27-28
  projected range, 26
  standard deviation, 26
Isolation, 217-218

Latchup, 199-200
  holding current, 213-214
  initiating current, 200, 213-214
  path, 200
LDD (see lightly-doped drain)
Lightly-doped drain, 138, 195,
                    239-259
  processing, 242-244
  simulation, 195, 242-244
Linear transconductance, 245
LOCOS, 218-223
  modified, 223-227

Maxwell-Boltzmann statistics, 84
Minimum overlap device, 258
Mobility
  electron, 135, 262
  electron mobility model
            in SIFCOD, 100
  Gummel's bulk mobility
            model, 89
  hole mobility model in
            SIFCOD, 100
  modified Gummel's mobility
            model, 89
  reduction, 268
MOSFET, 1, 2
  depletion mode, 271-279
  enhancement mode, 262-271
Multilayer coupling, 288-289

N-well
  CMOS, 200
  dose, 124

Narrow width effect, 201, 217,
                    222-223
Navier-Stock's equation, 55
Normal electric field, 89
Numerical simulation system
  block diagram, 17
  implementation scheme, 15-16
  system support, 16

Oxide
  incompressibility, 56
  thining effect, 53
Oxide encroachment
  LOCOS, 218-221
  SWAMI, 230

Packing density, 200
Parallel electric field, 89
Parasitic bipolar transistors, 199-200
Parasitics, 283-285
  extraction, 284-300
Pearson distribution, 27
Performance, transistor, 138, 245
PISCES II, 14
Poisson equation, 2-D
  in CADDET, 84
  in FCAP2, 113
  in GEMINI, 69
  in SIFCOD, 98
Polysilicon gate
  n+ 175-176
  over field capacitance, 293-296
  p+, 175
Potential barrier height, 161

Potential profile
  n-channel, 141, 147, 192-193
  p-channel, 180-183
  trench structure, 205-212
Power supply voltage, 145, 261
Process simulation, history, 14-15
Program interface, 16
Punchthrough, 149, 159
  bulk, 163-167
  current, 149
  p-channel, 181, 184
  point, 166
  surface, 163-167
  voltage, 149, 161, 270

Quasi-Fermi level, 147

Recombination
  Auger, 100
  Hall-Schockley-Read, 100
Reliability, MOSFET, 195-196, 240
Residual current, 132
  p-channel, 186

Saddle point, 163
Saturation region, 135
Saturation transconductance, 245
Saturation velocity, 135, 172
Scaling, 2-3, 261-282
  constant field, 2
  constant voltage, 3
  history, 1
  interconnections, 283
Series resistance, 245, 262

Short channel effects, 145, 155,
                          175, 271
SIFCOD
  basic equation, 98-99
  enhancement, 97-98
  flow chart 102-105
  input file, 109
  input format, 107-110
  output, 109
  program organization, 101
Simulation
  basic techniques, 131, 139, 144
  hierarchical, 15
  history, 13-14
  methodologies, 123-130
  system, 15-17, 286
  tools, 123
SOAP
  boundary conditions, 56
  flow chart, 57
  input file, 59
  input format, 58,61
  output plot, 60
  SWAMI simulation, 230
  2-D oxidation model, 54-56
Ston's method, 86
Stress, 230
Stress-relief oxide, 219
Submicron CMOS, 171
Substrate current, 240-241, 251-258
Subthreshold characteristics,
                          132-133, 147
Subthreshold leakage, 132, 186
Subthreshold slope, 132

Subthreshold slope (cont.)
  n-channel, 132
  p-channel, 184
Submicron CMOS, 171
Successive Line Over Relaxation
                 (SLOR), 86
SUPRA
  contour plot, 51-52
  device simulation, 141, 221-222
  diffusion simulation, 297
  fully-recessed oxide isolation
                 simulation, 48-49
  grid, 41
  input file, 50-52
  input format, 48-49
  LOCOS simulation, 219, 225
  SWAMI simulation, 231
SUPREM
  basic structure, 26
  device simulation, 139
  input file, 12, 38
  input format, 12
  output, 36, 37, 39
SWAMI, 228-229

TECAP2, 19
Thermal Oxidation
  linear-parabolic oxide-growth
                 model, 28
  non-planar oxidation, 53
  thin oxide, 27
  two-dimensional oxidation model,
                 model, 54-56
  volume expansion, 55
Thornbes's scaling rule, 89

Threshold current, 148
Threshold voltage, 132, 135,
                 148, 268, 275
  analytical calculation, 176
  depletion mode, 275
  simulation, 144, 147-148
  vs. channel length, 186, 191, 193
  vs. channel width, 223
Transconductance, 172
  linear, 245
  saturation, 245
Transistor
  current drive, 138, 172
  performance, 138, 245
Trench, 199
  CMOS, 199
  surface inversion, 201-215

Viscosity, 54
Voltage standard, 6

About the Authors

Kit M. Cham was born in Hong Kong. He received the B.S. degree in physics from Oregon State University in 1974, and the M.S. and Ph.D. degrees in MOSFET device physics from the Department of Engineering and Applied Science, Yale University, New Haven, CT, in 1977 and 1980, respectively. Since 1980, he has been with the Integrated Circuit Laboratory of Hewlett-Packard, Palo Alto, CA. He has worked on the characterization of MNOS devices for nonvolatile memories from 1980 to 1981. Since then he has been involved in the development of submicron CMOS technology. He is currently project leader for submicron CMOS device design. He is author or coauthor of 23 papers in journals and conference proceedings.

Soo-Young Oh received his Ph.D. in Electrical Engineering from Stanford University in 1980. He then joined Hewlett-Packard, and is now project leader for MOS device and processing modeling in the Integrated Circuit Structure Research Laboratory. He is currently working on a complete MOS two-dimensional simulation system from process down to circuit performance. His work has resulted in 15 publications.

Daeje Chin was born in Korea in 1952. He received the B.S. degree from Seoul National University, Korea in 1974, the M.S. from University of Massachusetts in 1979, and the Ph.D. in electrical engineering from Stanford University in 1983. He developed the two-dimensional process simulation programs SUPRA and SOAP, for his dissertation at Stanford while working part-time at Hewlett-Packard I.C. Laboratory.
After obtaining his doctorate, he joined IBM Watson Research Center in New York. He has been working on submicron memory processing technology and more recently high performance DRAM circuit design. He has 15 publications and four patents pending.

John L. Moll received his Ph.D. in Electrical Engineering from Ohio State University in 1952. He is currently senior scientist and manager of the Integrated Circuit Structure Research Laboratory of Hewlett-Packard. His extensive work in solid state devices has resulted in over 100 papers, ten patents, and a book, "Physics of Semiconductors". He is a fellow of IEEE, member of the American Physical Society, National Academy of Engineering, and Sigma Xi.